Praise for *The Truth A* *Rules of Busines*

"*The Truth About the New Rules of Business Writing* is filled with insights into improving all your written business communications. From everyday letters, e-mails, and reports to Web site content, proposals, and new media, you'll discover what it takes to make every written word count. A valuable resource, written in a concise and easy-to-read format, that you'll turn to again and again."

—**Jerry Allocca**, President, CORE Interactive

"An *Elements of Style* for our time...accessible, step-wise advice for communicating ideas to modern readers, whether of e-mails, promotional materials, or the printed page."

—**Susan Gilbert**, President, Interactive Elements Inc.

"The essential skill of today's businessperson is the ability to communicate clearly and persuasively. We are now often judged solely on the basis of our written communications. *The Truth About the New Rules of Business Writing* will make you a much better communicator. It's packed with examples, checklists, and easy-to-follow guides. This book has plenty of smart ideas you'll immediately apply to e-mails, letters, reports, Web, and social media. If you want to instantly improve the quality of your writing, this is the book to read."

—**Victor Urbach**, President, The Optran Group; Publisher, *The Urbach Letter*

"Times have changed. Business has changed. *The Truth About the New Rules of Business Writing* really does speak the truth for today's marketplace. It covers all the ground, including things like e-mail and social media. It should be required reading for people starting careers, looking to move up the ladder, or even when they start a business. It will be required reading for my sales team!"

—**Jim Josephson**, Vice President, U.S. Sales, Energy Advantage Inc.

"Filled with clear, practical principles and examples, *The Truth About the New Rules of Business Writing* is an essential guidebook for businesspeople who want their writing to get results. This useful reference book takes the fear out of putting fingers to the keyboard. Keep it on your desk because you'll refer to it often."

—**Julie Freeman**, ABC, President, International Association of Business Communicators

"All my clients are getting a copy of *The Truth About the New Rules of Business Writing*. It's packed with practical tips that even the most experienced writer can use. It's an easy read, but don't let that fool you. My new mantra is, 'Is it say-able?'"

—**Karen Susman**, The Networking Toolbox™, Speaker, Trainer, and Author

"Business is about communication, and Natalie and Claire have given us a crucial and invaluable tool. Whether you're writing a simple e-mail or a multimillion dollar proposal, *The Truth About the New Rules of Business Writing* is a *must read*!"

—**Jeff Goldberg**, Professional Speaker, Trainer, Coach, and Author

"The authors of *The Truth About the New Rules of Business Writing* understand that your true business abilities are transmitted to your colleagues and clients through your e-mails, memos, documents, and reports. If your writing is drab and old-fashioned, confusing or conflicted, you're in need of a writing makeover. This sharp, concise, and useful book gives you a step-by-step guide to making your writing shine with the clarity and impact that's demanded in today's competitive business environment."

—**Deborah K. Herman**, Founder and Publisher, *Building Long Island*

"Good writing starts with good thinking. This book shows you how to do both—and in a way that's effective for our electronic present."

—**Peter Krass**, President, Petros Consulting LLC

"*The Truth About the New Rules of Business Writing* is a masterfully crafted guide of best practices for effective and practical business writing in the 21st century. A dynamic writing duo, Canavor and Meirowitz provide an eloquent communication tool for print and digital media in today's evolving, fast-paced global society."

—**Jessica McAleer Decatur**, Director of Public Relations,
St. Joseph's College, Long Island Campus

THE TRUTH ABOUT

THE NEW RULES OF BUSINESS WRITING

Natalie Canavor and
Claire Meirowitz

Publishing as FT Press
Upper Saddle River, New Jersey 07458

FT Press offers excellent discounts on this book when ordered in quantity for bulk purchases or special sales. For more information, please contact U.S. Corporate and Government Sales, 1-800-382-3419, corpsales@pearsontechgroup.com. For sales outside the U.S., please contact International Sales at international@pearson.com.

Printed in the United States of America

First Printing December 2009

ISBN-10 0-13-715315-5
ISBN-13 978-0-13-715315-2

Pearson Education LTD.
Pearson Education Australia PTY, Limited.
Pearson Education Singapore, Pte. Ltd.
Pearson Education North Asia, Ltd.
Pearson Education Canada, Ltd.
Pearson Educación de Mexico, S.A. de C.V.
Pearson Education—Japan
Pearson Education Malaysia, Pte. Ltd.

Library of Congress Cataloging-in-Publication Data

Canavor, Natalie

The truth about the new rules of business writing / Natalie Canavor, Claire Meirowitz.

p. cm.

ISBN 978-0-13-715315-2 (pbk. : alk. paper) 1. Business writing. 2. Business communication. I. Meirowitz, Claire. II. Title.

HF5718.3.C366 2009

651.7'4--dc22

2009007022

This product is printed digitally on demand.

Vice President, Publisher
Tim Moore

Associate Publisher and Director of Marketing
Amy Neidlinger

Acquisitions Editor
Jennifer Simon

Editorial Assistant
Pamela Boland

Operations Manager
Gina Kanouse

Senior Marketing Manager
Julie Phifer

Publicity Manager
Laura Czaja

Assistant Marketing Manager
Megan Colvin

Cover Designer
Sandra Schroeder

Managing Editor
Kristy Hart

Project Editor
Betsy Harris

Copy Editor
Keith Cline

Proofreader
Water Crest Publishing

Senior Compositor
Gloria Schurick

Manufacturing Buyer
Dan Uhrig

Natalie Canavor

To my always supportive husband, Greg,
and to my daughter Victoria, always an inspiration.

Claire Meirowitz

Dedicated to my loving husband, Joe, and my wonderful
daughters, Diane, Laura, and Linda.

Note about the online-only content

You can access more materials—including "References," "Creating a style sheet," and "Resources"—by registering your book at www.ftpress.com/title/9780137153152.

Foreword: A new way to think about writing

Do you wonder why your messages don't work the way you want, and whether there's a better way to write them? There is. And we're going to show you exactly how to do it.

This book's goal is to make you a good business writer.

What do we mean by *business writer*? Not someone who earns a living as a writer, but anyone who writes for business purposes. You might turn out e-mails, letters, and reports as part of your everyday work in the corporate, nonprofit, or government worlds. You might be an entrepreneur or a professional—perhaps a lawyer, accountant, educator, scientist, artist—who writes proposals or articles as well as e-mails and marketing materials and blogs.

Whoever you are, writing can give you a powerful competitive edge. It can help you achieve your goals, work more efficiently, and persuade other people to your viewpoint. Even more: Good writing will help you be a better leader.

The strategies that *The Truth About the New Rules of Business Writing* demonstrates are based on the best ideas about practical writing, recast to work in today's fast-paced digitized world. Learn these ideas, and you're ready to write successfully not only for today's print and electronic media, but for whatever new communication vehicles emerge in our evolving business world.

On the Web

Visit this book's Web site at www.ftpress.com/title/9780137153152 to register this book and access bonus material, including reference sections on formatting, style sheets, and resources.

Acknowledgments

The authors gratefully acknowledge Neil Salkind of Studio B for his empathy, steady support, and publishing savvy.

And we thank our editors at Pearson—Jennifer Simon, who understood our intent and made this book work, and the whole editorial team that helped create a better book.

And, thank you, International Association of Business Communicators. This association brought us together, gave us the means to sharpen our thinking, and provided the forum for our "Working Words" columns, which led us to write this book.

About the Authors

Natalie Canavor has worked as a journalist, magazine editor, corporate communicator, and copywriter. She is currently a communications consultant and business writer whose publications, articles, Web sites, and video scripts earn international awards. Her features have appeared in a wide range of print and online media, including *The New York Times*. Earlier she created four national magazine start-ups and served as executive director of communications for New York State's largest educational agency. She also authored a successful book on marketing for photographers and leads workshops on writing for results.

Claire Meirowitz, principal of Professional Editing Services, manages projects and edits print, e-letters, and Web materials for a nationwide client base in information technology and banking, among others. She is also a writer and publication editor who cofounded and edited several journals in the labor relations field and award-winning newsletters for the education industry. She formerly directed publication and community relations for SUNY College at Old Westbury, where she oversaw production of more than 400 publications annually.

Together, Natalie and Claire own C&M Business Writing Services, providing writing, editing, and project management to organizations. They also present business writing workshops and write a column on writing and editing for the International Association of Business Communicators' online magazine, distributed internationally.

Introduction

Good Writing: What's in It for You?

Good writing is a big advantage in today's business world. You know that or you wouldn't be reading this book. But much more than you might imagine, effective writing can be your personal key to success.

Look around your own work environment: You'll likely find that good writers tend to get promoted, whatever their field. You may also observe that many business leaders write powerfully. That's not an accident—writing is a leadership skill, and can be just as important to success as good face-to-face skills.

When does good writing make a difference? *Always.* Every message you send matters.

Good writing is the road to more credibility, better professional image, and successful results. That can translate into more clients, enhanced relationships, bigger profits, and promotions. Because more and more of our work is accomplished through writing rather than in person, we depend on our writing skills to persuade, collaborate, manage, and lead.

Good writing works. In fact, this is a useful way to define good writing in the business environment: It accomplishes what you want it to do.

So how can you gain the advantage of writing well, a skill that few of us were lucky enough to learn in school or on the job? Use the tools we're sharing with you and absorb the ideas. We promise this will transform your writing, and your attitude toward writing.

Our decades as professional writers and workshop presenters have taught us that most intelligent people need only two things to write successfully:

- First, you need a clear idea of what good writing looks like and sounds like *today*—not according to outdated 20th-century models.
- Second, you need a system to think through any writing challenge you're faced with, from e-mails to proposals, reports, Web site copy, blogs, and much more. Based on our combined experience

as journalists, corporate communicators, and editors, we've created a step-by-step strategy that takes you from the very beginning to the end, guiding you to produce a piece of writing that accomplishes exactly what you want—without any agony.

This book is written expressly for adults in the business world, and it's based on methods that have nothing to do with traditional ways of teaching.

Here's the inside story on good writing: You don't have to spend days or weeks or years drilling on grammar, punctuation, and spelling. If you want to play baseball, mastering the rules is useless if you can't catch the ball or hit it. Writing is not about rules either—essentially it's about thinking.

The importance of good writing isn't something that gets talked about a lot. Most people don't notice the quality of the writing they see—they simply react positively, negatively, or not at all. They may not read the piece, or may skim it. They may fail to understand the message, or find it unpersuasive.

That's why there is growing recognition that writing is an essential business skill. Poorly written communications fail to explain a company's products, values, and messages—or worse, describe them in a negative manner. Competitive advantage is lost, and so are opportunities to connect with customers, colleagues, and collaborators. Business efficiency is sacrificed.

Don't think American industry hasn't noticed. A few years ago, a blue-ribbon group called the National Writing Commission surveyed top business leaders and found that poor employee writing is such a big problem that companies spend more than $3 billion per year trying to improve it. Their report was called, "Writing: A Ticket to Work...Or a Ticket Out," and documented the importance of writing in decisions to hire and promote, especially in high-growth industries.

What does that mean to you? *Opportunity.* We invite you to take advantage of all the writing experience and expertise that we've built into these pages. They'll equip you to write powerfully for today's business media and help you succeed in today's competitive business environment.

TRUTH 1

Most people aim for the wrong target

There's an old saying, "If you don't know where you're going, any road will take you there." Or, if you like sports analogies better: "It's much easier to hit a target you can see."

Most of us never learned to write for practical purposes. Early in our school years we were drilled on mastering grammar and punctuation, and in high school and college, usually practiced a literary style suitable for academic essays.

Then we enter the business world, and unless we're lucky enough to find a writing mentor, we encounter few good models to light the way. Most people write carelessly, using old-fashioned, outdated styles that are ill suited to today's tempo and spirit. So we muddle along without knowing how to improve our own written communication.

Does this matter? Yes! Because it produces writing that doesn't work—meaning it fails to accomplish our goals.

So the first step in writing-for-results is to describe the characteristics of strong writing. Having a clear vision of the target is a huge part of getting there, and it enables you to apply skills you already own but may never have thought about.

Mastering rules and imitating outdated models will not make you a powerful writer. Understand what works and apply that knowledge to everything you write.

Define good writing yourself

We're sure you know more about good writing than you think. Try this: Make a list of everything you don't like about other people's writing, whether in books, newspapers, e-mails, letters, newsletters, on the Web, in blogs, and so on.

To start you off, here are some of the characteristics that come up on most lists when we do this as a group activity:

- Hard to understand
- Boring
- Confusing
- Illogical
- Wordy
- Message obscure
- Purpose unclear
- Hard to read aloud

Add as many characteristics to this list as you can—and you've defined bad writing! For more descriptors to expand your personal list, see the "bad writing" rundown later in this chapter, drawn from our workshop participants.

Next: What happens when these negative words or phrases are reversed to make them positive? The preceding words or phrases might reverse this way:

- Easy to understand
- Interesting
- Clear
- Logical
- Concise
- Message obvious
- Purpose obvious
- Easy to read aloud

Take your full expanded list of negatives and reverse them.

Now you've successfully defined good writing and know what to aim for. You need only to believe in your definition and that it works for every purpose. This alone will immediately put you way ahead of most people.

Notice that incorrect spelling, punctuation, and grammar might not even have come up on your "bad" list, or may have turned up at the end. Instinctively, people know that technical know-how doesn't define good writing, although it matters because technical mistakes interfere with clarity and understanding. Today your computer programs can give you plenty of help with the technicalities—we'll show you how to use that support effectively.

Bad writing: the whole sad story

- Annoying "style" that interferes with the message
- Ambiguous
- Doesn't hold your attention
- Can't tell why you should care
- Makes the reader spend time figuring out the meaning
- Can't tell at whom it's aimed
- Repetitive
- Uninteresting word choice
- No detail to back up ideas or generalizations
- Vague
- Dense—no breathing room for reader to absorb information
- Pompous

- Connections between ideas and facts not clear
- No rhythm
- Stilted and awkward
- Rambles
- Full of clichés
- Too much—or too little—information to make the point
- Makes leaps, without helping reader to follow
- Feels passive
- Unconvincing—full of empty claims, unpersuasive
- Full of jargon of any kind—academic, scientific, business, and so forth
- Doesn't deliver what it promised
- Bad grammar and structure
- Misspelled words
- Incorrect word use

Good writing: a definition to live by

- Invisible style that lets the message shine through
- Unambiguous
- Holds attention
- Makes it obvious why you should care
- Audience it's written for is obvious
- No repetitions—words, ideas, examples
- Interesting word choice
- Right level of detail
- Allows breathing space, easy to absorb
- Feels natural
- Has natural rhythm
- Flows smoothly
- Stays on point
- No clichés
- Right amount of information to support points
- Logical connections
- Feels active
- Persuasive
- No empty rhetoric—backs up all claims
- No jargon

TRUTH 2

If you can say it, you can write it

A colleague sent us a press release that began, "Transformation of a dream often begins with acts of imagination that elevate a starting vision of change above the intimidating presence of things as they are."

"Jack," we asked on the phone, "what do you mean?"

"Oh, that you've got to have a dream to make the dream come true."

Why didn't Jack just say what he meant, or find other simple words to say the same thing without sounding like Rodgers and Hammerstein? Why do we write so much "stuff" that is so different from what we'd say if someone asked us in conversation what the new gizmo does, or why the acquisition is so important, or how the new system works?

Business writing today often substitutes for conversation, so it works best when it's close to spoken language.

Maybe the answer comes from another famous theatrical production number: "Tradition." We may have learned to write in long, complicated, formal fashion to fill up English term papers or impress professors. In the business world, we may have learned to imitate other people's bad writing.

How to escape from this morass? By believing in this basic principle: *Good business writing today is very close to good spoken language. So, if you can say it... you can write it!*

We're aiming for you to write the way you speak: clearly, unpretentiously, using short words, short sentences, simple structures, and a natural flow of ideas. Historically, oral storytelling predated writing by eons, and most of the greatest classics in every language—the Bible, Homer, Chaucer—were written records of oral tradition.

But this spoken-word basis got lost somewhere after the printing press was invented in the fifteenth century. Now, happily for writers in the business environment, it's back. No one has time to write long-convoluted-dense-abstract-ambiguous messages...and almost no one wants to read them.

The definition of good writing we developed as part of Truth 1 is universal. Effective business writing in the twenty-first century has come to mean writing that is simple, direct, clear, easy to read, and...conversational. When it's read aloud, it sounds a lot like spoken communication.

Of course, there are important differences between spoken and written language, and you need to know how to make the transition from the first to the second—and we'll show you how to do it. But the important point to remember is that successful writing "sounds" a lot like speech. In fact, many professional writers routinely read their work aloud to be sure it is say-able. When it's not, they make changes.

"Old" business writing vs. "new"—Here's a kind of statement that is often seen in business writing. Actually, one of the authors wrote it for this book:

The amount of human interaction in contemporary office contexts is continually diminishing, because of technology.

When we checked out this sentence, we realized that it didn't meet the say-ability test; when read aloud, it's awkward and wordy and has a sing-songy tone. You'd never express the idea this way when talking to someone.

Here's a rewrite that's much closer to speech:

Today businesspeople don't talk to each other very much; we mostly send e-mails.

How "saying it" helps you connect—There are many reasons why a conversational approach to writing succeeds in the business setting. Personal contact is diminishing. More and more of us work out of our homes. Even in office situations, we may be distant from colleagues and those we report to, or who report to us. We connect with the outer world to an amazing degree through e-mail, letters, reports, proposals, Web sites, blogs, texting, online communities, and print materials.

The human race isn't really ready for this. We are accustomed to making judgments about people's trustworthiness, value, and simpatico level in face-to-face situations. Imagine holding a jury trial where all the information is delivered by e-mail and you never see the accused, the judge, lawyers, or other jurors. Would you want to make a decision only on the bare facts? Without hearing tone of voice, seeing expressions, and observing body language and interactions? Observing the pauses and hesitations?

The lesson for writing: You need to recognize what can and can't be accomplished in written communication and take account of those limits. But here's something that doesn't get a lot of attention: *When the written word is substituting for in-person interaction, a conversational style can be a big asset.*

Further, in using a spoken-word base to plan a communication, you give yourself *the advantage of suggesting individuality or even personality in your message.* Business communication need not be coldly factual and neutral. Much of it would work better if it seemed more individualized.

When you think about how you would say something—especially to a specific person—you automatically choose words, expressions, and structures that characterize a personal interaction. Your language picks up warmth and spontaneity that can be very effective. Of course, on the other hand, you have to guard against displaying negative feelings, which can happen with equal ease.

TRUTH 3

Forget yesterday—write for today

Your challenge: Write an e-mail message to the Planning Committee setting a time for an important meeting. Here are two ways you could do it:

Yesterday's way

On June 18th, 2009, at 2 p.m., a meeting has been scheduled to follow up on our previous conversation of May 10th, and I would ask you to make arrangements to be in attendance. It is anticipated that approximately two hours will be necessary to cover the agenda thoroughly. Please advise me of your availability.

Today's way

Can you come to a meeting next Thursday at 2 p.m.? I'd like to follow up on all the great ideas I heard when we talked a few weeks ago. I hope you can give this about two hours. Please e-mail back that you can be there.

Which is better? Why?—The traditionally written Yesterday's Way version is wordier, passive in tone, takes longer to read—and takes more time to write. Furthermore, it sets the meeting participants up to expect a dull old time.

Today's Way is much closer to what you'd actually say to someone: It's clear, spontaneous sounding, and enthusiastic. The writing is transparent, meaning that the reader notices the message, not the writing itself. In Yesterday's Way, how the message is written actually slows down understanding.

Also, framing the message as a personal contact prompted the writer to include more persuasive content about the event's nature. If you received this version, wouldn't you feel better about committing two hours of your time?

Traditional business communication is wordy, passive, and dull. To work in today's speed-charged, over-communicating world, it must feel crisp, fast, and transparent.

We know that many business meetings can't promise a sparkling agenda. But when you've got something positive to work with, flaunt it.

Try saying it—To see how the "saying it" idea can help, try this: Think about something you have to write about, like the answer to a question. It can be along the lines of, "What's important about Project X that I'm working on?" Or, "Why will my new idea about Y be an improvement over what we do now?"

Then say the answer—preferably aloud—the same way you would say it to a friend or colleague in person. Act as if it were a phone conversation with a particular person. Then write it down that way.

Next, look at what you've written. You may be surprised at how close you've come to a good written version of your statement, and how fast you accomplished it without any trouble at all.

What you wrote is probably not perfect. You might need to adjust it to conform to the written language, eliminate unneeded repetitions, or fix the grammar or structure. Or you may notice that what you've said is confusing or that something is obviously missing. This tells you that your message needs more thought, and that you must get your facts and ideas in order before writing.

The bottom line: You already possess the most important tool. Your own language and conversational skill give you the "say it" foundation that we recommend for writing.

There may even be a scientific basis for why this works. Researchers are finding that different parts of the brain function in highly specialized ways. For example, color and shape are processed separately and then coordinated. Written language and spoken language may be processed in different regions of the brain, and this would explain why most of us can be perfectly lucid when we talk about a subject, but become stilted and confusing when we write about the same thing.

So when you have a writing problem, see what happens if you try to talk about it, as if you were speaking to a specific person. You may find yourself much more articulate, and can then go on to develop a good written version. The approach works just as well with complicated documents, like proposals and reports, as it does with e-mail memos.

TRUTH 4

Planning is the magic ingredient

Most people we know spend either no time at all planning what to write, or agonize endlessly about where to start, what to include, how to organize the message, and so on. If you're in either category, we have a solution—really.

It's a step-by-step system that works for just about everything. Here's a quick overview:

1. Identify your goal. What's the message? Desired outcome?
2. Think about your audience. Who is it? What do you know about the person or group?
3. Determine the right tone and format, based on answers 1 and 2.
4. Map the content based on your answers.
5. Organize in a way that works for your purpose.
6. Begin with a strong lead.
7. Write the middle, quickly.
8. Build your conclusion.
9. Read what you've written and evaluate, cut, fix, tighten, and sharpen.

When you add this approach to your new definition of good writing and the write-it-like-you'd-say-it idea, you are equipped for power business writing.

Every message you write matters. Don't be fooled by how carelessly an e-mail can be sent—e-media demand even better thinking than print.

Why use this strategy?—Before we discuss these steps and demonstrate them in subsequent chapters, we want to answer a question you're probably asking: "Do I really need to make a project out of something like a simple e-mail?"

The answer, in a word: *Yes.* Because everything you write reflects you. If you scribble a quickie e-mail in all lowercase letters and no punctuation and with no thought to what it sounds like, well—we hate to break it to you—that's what people are going to think you're like: sloppy, disorganized, and unthinking.

Even if you're only scribbling fast notes to a colleague, you never know who's going to see them. E-mails are often kicked upstairs, and as some politicians and business leaders have learned the hard way, e-mails are indelible. Have you noticed how many political and business leaders are losing big time because their thoughtless e-mails—memorializing an unethical or illegal practice, for example—were unearthed years later?

> The less space you have to deliver your message, the better and clearer your thinking has to be.

People got used to writing shortsighted, ill-considered letters, memos, and other records in the days when an original and a few file copies could be contained. But you can't shred an e-mail or control its spread.

Even in the digital age, you need to plan—Does electronic delivery itself promote carelessness both in content and style? Yes, we think; the ease of composing and sending an e-mail or jotting a quick blog posting lulls us into assuming that a stream-of-consciousness approach works. And of course, we've replaced a lot of conversation with e-mail exchanges, adding to the delusion that unpremeditated writing is good.

The news is: Writing with e-media takes at least as much planning as with the old print systems. E-mailing and blogs look like spontaneous media, but they're not. In fact, as our digital world advances, messages have to be shorter and tighter than ever. Think of e-mail versus typed and copied interoffice memos (if you were working that far back), or text messaging versus e-mail now. The less space you have to deliver your message, the better and clearer your thinking has to be.

If "the carrot" approach doesn't motivate you to care about writing, consider "the stick"—the negative impact poor writing can have on your future. In the short run, you may lose job opportunities, assignments, references, credibility, effectiveness, or good relationships.

In the long run, who knows what you will achieve if you write to your advantage rather than disadvantage? Recent studies prove that writing skills get people hired in the most exciting fields... get them promoted...gain them leadership roles. We all know about people who've written great blogs that led to successful careers. Warren Buffett takes the trouble to write well, and so can you. Our strategies will give you everything you need.

Will our approach slow you down? You may or may not write faster right away. But definitely you'll write a whole lot better and more confidently. You won't agonize over stilted, old-fashioned ways of delivering your message. You won't worry about how to start or how to make your point, because you'll have a complete, simple-to-use, step-by-step guide to support you through every writing challenge that comes your way.

TRUTH 5

To achieve your goal, look below its surface

Consider Andrea. She's won an award and thinks it's a good idea to let her boss know. So she writes:

Dear Joe,

I thought you'd like to know that I've been chosen as my local Chamber of Commerce's Person of the Month. They're holding a lunch in my honor on October 5th.—Andrea

Communicating this honor is a good move, but did Andrea do it well? Not really. She could have gotten a lot more mileage from it than she did. There's a simple question she should have asked herself, and it's the question you should ask yourself every time you write anything beyond a three-word message: *Exactly what am I trying to accomplish with this piece of writing?*

Put it in words—articulate it. Don't write anything before you can do this, because a fuzzy purpose produces a fuzzy piece of writing. Knowing your goal makes a huge difference.

You may think that much of your writing doesn't have a defined purpose. This is a mistake. In the business world, almost all writing is created for a purpose. Most are pretty basic, and they overlap: to inform, share information, document, persuade to a point of view, sell something, motivate, inspire, and build relationships. Even thank you letters fulfill one or more of these goals. Often you need to accomplish several of these things at once.

On a more specific level, goals are usually very concrete: You might want your reader to come to a meeting, fill out a form correctly, buy into a project, authorize overtime, assign you a new project, or answer a question about your current work, for example.

> To make your writing work, define the specific purpose of each communication and always know your personal sub-agenda and build it in.

But that's not the whole story. On the surface, you're simply sharing a piece of information. That's what Andrea did in writing to tell her boss that she won an award.

What to get across—Let's stop and think. What, other than the plain facts, might Andrea want to get across in delivering her good news? She might (and should) consider that she has a chance to impress the supervisor with her increased value to the company. She could also take the occasion to personalize their relationship.

What happens if she defines these two possibilities as the goals? Instantly, ideas suggest themselves for the message's content. She could say, for example, that

- The award recognizes her professional position, as well as her work as a volunteer.
- She appreciates the opportunities the firm gives her professionally and the encouragement to do community service.
- She'll be giving a short acceptance speech mentioning the company, its products, and its contributions to the community.
- She hopes her boss will come to the luncheon as her guest, so that she can introduce him.

Even if her boss declines the invitation, she's way ahead: She's made herself more memorable. She's strengthened the relationship. He will probably send the message up the corporate ladder, because having a staff member honored and spreading the company word reflects well on him

When you look past the obvious to figure out your larger goals, the content of your message will almost always shift, sometimes radically. *Knowing what you want is essential to knowing how to get it.* And you'll see opportunities to achieve your personal long-term goals as well as the immediate one.

If Andrea defines her goal as demonstrating her enhanced value and improving relationships, she can write a memo that actively supports her career advancement.

This is her sub-agenda. You have one or more of your own—maybe you want to grow your business, get promoted, create a strong support network, or find a new job. And once you take your personal sub-agendas into account, you'll begin to see writing not as a nuisance, but a great opportunity to showcase your professionalism, credibility, competence, creativity, loyalty—whatever.

Notice, too, that once Andrea has clearly identified her goal, the actual writing almost takes care of itself. Moving from the checklist to the finished message becomes really easy. Here's one way she could shape the message as an e-mail.

Subject: Good news and an invitation

Dear Pete:

I'm happy to tell you that the Fairmont Chamber of Commerce has chosen me as Mover & Shaker of the Month, in recognition of both my professional standing and contributions to the community. I'll receive the award at a luncheon on Wednesday, November 12, and I very much hope you can be there as my guest.

I'll have an opportunity to speak briefly. I plan to talk about the company, and especially, how management supports community service. The Good Neighbor Program gives me wonderful opportunities to help local charities, and I welcome the chance to express my appreciation publicly.

Can you attend? I'd love to introduce you to the group.—Andrea

Did it take more time to write this message compared to the one Andrea planned on? Of course...but think of the payoff. Beyond what she accomplishes for her own positioning, Andrea is able to reinforce a company policy from which she and others benefit. Learning that an employee is being honored for volunteer work, and will therefore deliver a positive company message to a significant audience, can only make management happy and confirm the program's value.

TRUTH 6

Cut to the chase: Put the bottom line on top

Suppose you're reporting on the results of an extended project you supervised, whose purpose was to investigate a potential new workflow system. Your basic goal is to inform and document. But that's only part of the agenda. As in many situations, you want to accomplish much more. For example:

Goals checklist

- Tell your boss that the project was a good investment and that his/her faith in your ability as a manager was justified.
- Let readers know that the results are significant.
- Demonstrate that you've done a careful, thorough job on the project.
- Make your interpretation or conclusions credible.
- Present your work in the way that's clearest, most logical, and, if possible, interesting.
- Persuade the rest of the organization or the higher-ups to act on your results.

If you articulate your goals in this way, think about how your content choice is affected—and *content is always the heart of effective writing.*

Most e-mail memos contain much more information than they need. Articulate your goals and you'll recognize exactly what content to include.

Content checklist

- Include a clear statement about why the project was carried out and its importance.
- Indicate exactly what the results were and their significance.
- Outline or narrate what you did and how you did it.
- Marshal the evidence that supports your conclusions.
- Indicate who your project team or collaborators were and any information about time frame and budget that's relevant.
- Clearly state the course of action you recommend, and the benefits that may be expected.

In this situation, the order of the content checklist works. But often you may want to juggle. Be sure to establish why readers should care, and the results, right up front.

Almost always in business writing, put the bottom line on top.

Look at our two lists—goals and content—and you'll see that by articulating your goals and then figuring out how they determine content, you've created an outline (with no agony at all).

Check out the reasoning in the following examples. We think your approach to memos and e-mails will change.

Challenge 1: Write an e-mail memo to your boss about your project needs.

Yesterday's way

Dear Ms. Smith:

As you know from our previous meetings and earlier discussions, the Alliman Project has a major deadline that will be coming due in three months. To preserve the project milestones, our office will need to purchase a variety of tools and products at this time. I am including a list of project needs below, and am asking you via this memorandum to authorize these, so that the Purchasing Department will consider them legitimate company purchases.

Sincerely, Tom

Goal analysis: You want a swift, affirmative response without provoking objections that could cause delay. So you need to say just enough for your message to be read quickly and instantly understood. You want to sound matter of fact, businesslike, and efficient. Combining this goal with the conversational approach would give you something like the following version.

Today's way

Hello Jessica,

Please sign the attached list of project needs so that Purchasing will buy them for us.

As you know, the deadline on the Alliman Project is August 1, so I need your authorization quickly for these necessary tools and products.

Thanks for your help.—Tom

The difference: The "call to action" is up front and center. The boss can't miss it; whereas in the traditional version, she'd have to search for it. Because you first articulated your goal, you didn't include all the extraneous information. And, the old-style memo is much more likely to generate an inquiry about how the project is progressing and whether the purchases are justified. Also, Tom comes across as efficient, effective, and reliable, so the memo contributes to advancing his sub-agenda.

Challenge 2: Your vice president has asked for a formal written update on your project.

Yesterday's way

Enterprise Project Report

a. Overview

b. Timeline

c. Products required; availability

d. Staffing needs; new hires

e. Sites

f. Progress toward goal

g. Steps required to complete project by August 1

h. Anticipated completion date with and without necessary resources

Goal analysis: You want the VP to quickly see what's needed and support you. How to accomplish this? A lengthy, detailed document going over the same old ground, with the progress section buried, as in the outline above? Unlikely. Ask yourself, what does she really want to know? What's important for the company? What highlights your accomplishments? This should lead you to:

Today's way

Enterprise Project Report

a. Progress toward goal

b. Steps needed to complete project by Aug. 1, including consequences if necessary resources are not provided

c. Brief project recap, including timeline, personnel, products, and sites

The difference: Because you defined your goals first, this report is likely to be read and is likely to be passed up the chain of command to people who can provide the resources you need. What's important is succinctly stated, right up front, and those who want detail will find it in the recap.

Notice that often the more closely you define your goal and articulate what you want to accomplish, the less you have to include—because you've figured out what counts and can focus precisely on it. That means less writing.

TRUTH 7

"Me"-focused messages fail

"Who's the target?" That's a question every professional writer working in a business environment asks before writing anything, right along with, "What's my goal?" The target is the intended audience, the one person or many people you're addressing. Does it make a difference?

You bet. So: Define your audience and analyze it as completely as you can.

Let's assume that you're writing to an individual, or to a few people. What can determine how they will receive your message—and therefore influence how you shape it?

It depends on the particular situation, but you may need to take factors such as the following into account:

- Age/generation
- Gender
- Educational level
- Role in the organization
- Your relationship to the persons (Peers? Superiors? People who report to you?)
- Strong suits (Technical? Relationships? Detail oriented? Big-picture thinkers?)
- Weak suits (Technology challenged? Impatient? Short attention span?)
- Cultural/ethnic/religious background
- Language (Native English speakers or a global audience?)
- Their interests and what they care about
- Their hot buttons
- Your competition for their attention
- Their preexisting beliefs about whatever you're trying to pitch

Take the time to figure out what your recipients care about, who they are, how they see the world. Frame your message based on these factors.

Only some of these factors are usually relevant to a particular writing situation. But you need to understand the target audience's viewpoint as well as you can and take it into account. *You can't communicate successfully if the message doesn't reach the receiver in his or her own language and terms.*

Many communications fail because the message senders are so focused on what they want to get across or accomplish that they almost totally ignore the receivers.

There's an old lawyer's axiom that says: Never ask a question you don't know the answer to. *For writers, the rule is: Always know how you want the reader to react.* To do that, know who they are.

Primary and secondary audiences—Professional writers think in terms of primary audiences and secondary audiences, and it's a useful concept. *Primary* are those you're directly addressing. *Secondary* audiences can include others who may also end up seeing your message. You report in writing to your boss, and he or she reports to his or hers—perhaps incorporating what you sent him—and so on up the line. Your innocent e-mail can wind up in the organizational stratosphere. So can e-mails you send to a colleague or buddy.

But, we hear you saying, how can I run through a list of audience characteristics like this one every time I write an e-mail or letter? To do that, we have some terrific shortcuts for you.

Quick audience visualizing trick—To get on the same wavelength as your target audience, visualize that individual in your mind.

Take the time to really see him or her: facial expressions, gestures, clothing, stance, mannerisms; hear the person's voice talking to you, maybe even laughing. Automatically, your writing will adopt the right tone because you're triggering the same instincts that guide you in face-to-face conversation. Your writing becomes almost an interaction, tailored to that target.

If you're writing to a prospective customer you've met, for example, visualize that customer in as much detail as you can and ask yourself, what would this person want to know? What would his/her questions or objections be? What would show this customer that I understand the company's problems and can offer solutions?

Once you've brought the person alive in your mind, visualize a conversation with him or her on the subject you're writing about. Doing so will often provide you with important clues to how that person would respond in a face-to-face situation, and you can take account of this in shaping your message.

What if you're writing to a group? Think individual anyway. Pick a "typical" member of the audience to visualize. If you don't know one, think of someone you know well who could be a recipient. Warren Buffett, the financier famous for his clear communications, was asked how he wrote so well about complicated subjects. I always think of my two sisters, he reportedly said, who are very intelligent but not sophisticated about finance, and write to them.

See the world through other people's eyes—Here's another way to cut to the chase. Put yourself in your target audience's shoes and answer this one question: What's in it for me? Thinking about this before you write (or speak) will always make you a better communicator. This is the WIIFM principle referred to throughout this book.

Ask why they should care. What do they have to gain or lose? How will they be affected? Employee communication provides a good example of why a whole program can succeed or fail based on whether the WIIFM question is posed. A company cannot gain support for a change in health benefits by explaining how much money management will save. Every employee will want to know exactly how he or she is affected. If the effect is adverse, it may be important to explain what trade-offs were involved, again in personal terms (for example, a higher deductible is necessary to counter rising insurance costs while maintaining good health benefits).

How messages are delivered—If you've attended a workshop on interpersonal communication, you may have heard the presenter cite a study about message delivery. The trainer asks: In sending a message, what do you think is the relative importance of body language, tone of voice, and words?

Unless audience members have been in similar workshops before, most are astounded to learn that 55% is body language; 38% is tone of voice; and only 7% is words.

The study was done by psychologist Albert Mehrabian back in 1971. It tells us something important about writing: When we communicate through the written word, we're missing tone of voice and body language altogether. This absence can make it hard to know what the writer actually means. Using sarcasm in an e-mail, for example, or a joke, is risky because with only the written words to bear the message, your meaning is open to interpretation.

TRUTH 8

People are not the same: Write for differences

As part of a business writing workshop for new managers, we asked the participants to write a memo asking their supervisors for time off during an especially busy week. Many in the group had a lot of fun with the request, inventing caretaker responsibilities, a daughter's wedding, a sweepstakes cruise offer, and so forth. Most of the writers did a good job of stating their cases, but they all would have failed: Not one mentioned how the writer's work responsibilities would be covered during his or her absence.

They should have put themselves in their supervisors' shoes and asked, "What's in it for me—or not in it for me? Will I be inconvenienced? Have to work longer hours myself to get the work done? Pay for extra help?" The supervisors would have rejected the requests or, at best, asked, "What about the work?"

To accomplish your goal, think in reverse: How will my message affect the person or group? What's in it for them? What resistance should I expect?

Every time you write, pause and answer the "What's in it for me?" (WIIFM) question if you want to succeed. An innovative media artist we know failed to do that when writing to several research labs, asking for access to neurological testing equipment. She described her video art and art-world credentials well, but her proposal failed to explain what was in it for the labs. Once she had thought through what she could offer them and how they might gain by collaborating, she made a good case—and was successful.

When you're writing to accomplish a goal, knowing your audience and seeing things from their perspective tells you what content will work. If your company is changing how it calculates retirement credit, for example, each different employee group might need a separate answer to the WIIFM question. If you realize that, you're one giant step ahead of the game.

Be aware of generation gaps—Dara, a 30-ish assistant marketing manager, was asked by her 40-something boss, Melanie, to attend an industry event and report back on it. Dara wrote:

Hi Mel—the meeting yestrday was like totally b-o-r-i-n-g. Guys in suits dulling us to death with industry trands. But I gave out quite a lot of business cards. do you want me to go to the ABD meeting next Tuesday.

Will Dara go the ABD meeting? Has she impressed her boss?

Way beyond the obvious technical mistakes, Dara failed to consider her goals and her audience. Thinking about goals would have told her that 1) she had a chance to perform well in person and that 2) in writing she had to deliver what Melanie wanted, if she is to receive more opportunities.

Thinking about audience would have told her that she should not address Melanie as a peer. Your workplace probably has a mix of generations, and much has been written about the differences between them:

- **Generation Y:** Born after 1982...youngest members of the workforce characterized by energy and huge ambition to succeed quickly. High degree of social consciousness and confidence with technology; global outlook; minimal institutional allegiance.
- **Generation X:** Born between 1965 and 1981...typically seen as "challengers," who want to find better ways of doing things and change the rules and processes. They are prepared to work hard and expect to succeed. They tend to like teamwork.
- **Boomers:** Born between 1945 and 1964...come with high expectations and are driven to high levels of performance. Generally have an individualistic, sometimes self-important outlook and often an authoritarian approach to power.

Where do most of today's leaders fall? The boomers are still very much with us. Most of those you report to may be only somewhat older than you, but the significant higher-ups are probably boomers. They may not represent your primary audience, but very often they represent your secondary audience—the people beyond your immediate super-visor to whom your report or e-mail may be relayed, to your credit or discredit.

As a general rule, the older people are, the more formality they prefer in their lives and in communication. They will probably not appreciate e-mails with smiley faces, text message abbreviations, or a super-casual style. They are unlikely to respond positively to sloppy writing with bad spelling, haphazard punctuation, and no capitalization.

The higher-ups in Dara's company are probably boomers, and if her e-mail climbs into their territory, she's made an even worse impression than was immediately apparent.

What if you report to someone younger than yourself? Treat young supervisors as if they are older, as a sign of respect. Young people in high positions typically need to feel respected, even when their manner is casual. And they can be very sensitive to signs that respect is lacking.

What Dara should have written—Seeing the situation through Melanie's eyes would have led her to include the following:

- Who was at the meeting (total number, notables)
- Whom she spoke with and any relevant business exchange
- A nutshell account of the presentation, noting points relating to company interests
- Appreciation for the opportunity to attend

This content might be a bit heavy for an e-mail, so Dara could make the report an attachment (unless the recipient is attachment phobic, something necessary to know). The cover memo might read:

Melanie, here's a report on the LJA meeting you asked me to attend.

I enjoyed the opportunity to learn more about our industry and meet some new people. Let me know when you'd like me to do something like this again.—Dara

The report itself? She can now do a businesslike version that follows the content list. She can add a heading, her name, and use subheads for each section: Participants; Nutshell Panel Discussion; Relevant Points; Contacts Made.

What has Dara accomplished if she does it this way?

She's shown her boss, and probably her boss's boss, that she is interested in the business, able to make the most of an opportunity, able to communicate in writing, and able to represent the company well. She has shown that she is trustworthy, discerning, and a good candidate for more responsibility.

Can something as simple as writing a good report do all that? Absolutely.

TRUTH

Tone makes—or breaks—your message

One day recently, Tom received an e-mail from a man named Peter, who wanted Tom's company to hire him to give a presentation for the Human Resources department. It began, "Hey there! I'll bet your guys haven't seen this kind of show before, and you'll be a hero for bringing me in." The problem was that Tom's company is a staid law firm with many of the partners and associates in their forties and higher. What was jarring about Peter's message? Its tone, based on the obvious fact that Peter hadn't researched his target audience.

When you're writing something that is simple and not too important, you may only need to consider "goal," "audience," and "tone" for a minute. But when your purpose does matter, even a short message should be carefully plotted out. If you're about to write something substantial, such as a proposal, get your goal and audience ideas in place with care, and the rest will follow far more easily than you'd expect.

Deciding on tone—Think of tone as the general atmosphere and "feel" of your message: Formal? Casual? Friendly? Breezy? Enthusiastic? Motivational? Inspirational?

These decisions will automatically trigger your choice of words and writing style. So how do you know what tone to use? It's usually a logical outcome of goal and audience.

Relationships should guide the choice of tone. The voice you use to address a top executive, a colleague, a subordinate, or a client is critical, and different in each case.

Let's look at a fairly simple message that's important to the writer. Marian needs to call a meeting that will bring together her own boss, someone who reports to her, and a colleague on her own level. Her goal is clear: getting everyone to the meeting ready to play their roles well and advance the project she's running.

Marian's high degree of investment means she should think things through carefully. She has three distinctly different relationships with her recipients. So one memo will not work in this case. And the different relationships suggest different tones.

When you write to your supervisor, you naturally write in a different manner than you would to someone who reports to you, or to a friend. Marian needs three customized memos.

To her subordinate:

Dear Mark:

The Project Win meeting is scheduled for Thursday, May 12, at 2 p.m. As planned, I will call on you to deliver the Section A progress report, so please be prepared and bring 10 copies of your handout. If you have any problems developing the report, be sure to let me know ASAP. In any case, let's go over the gist of your report and the handout together by Monday. Please call Martha to set a time.

To her supervisor:

Dear Alpha:

I've planned the Project Win meeting you expressed interest in for next Thursday, May 12. I consulted with Jane about the date, but the time can be adjusted to your schedule. Is 2 p.m. convenient for you? If so, we can begin with the progress report. Or, we can plan that for 2:30 or 3 if that works better. Let me know as your schedule shapes up—thanks.

To her colleague:

Dear Joan:

Can you come to a meeting about Project Win on May 12 at 2 p.m.? A progress report is on the agenda, and I know you want to hear how my department is doing with that. Alpha will join us for part of the meeting. Let me know if you'll be there, and if you have any concerns—let's talk about that before the 12th. Maybe lunch?

How do these memos differ? Obviously, the content had to be adjusted for each recipient, but also the tone (the sound of the "voice"). Which is the least formal? The message to your peer—a colleague and equal. Just as in person-to-person contact, actually. Your communication to a supervisor demands a relative formality and should show consideration for her demanding schedule; someone reporting to you also needs formality and must be presented with clear expectations. But you can relax somewhat with your peer, and adopt a more collegial tone, though not so much that you overlook what's needed to reinforce mutual support.

What's your guide?—Relationships are your guiding force to shape tone. Consider your standing with the person, add the audience-analysis techniques covered in Truth 8 to tailor your approach to the individual and address his or her self-interest, and you're sure to get it right. The visualizing trick, where you focus on seeing the person in your head, can be extremely helpful.

Some "nevers" (well, hardly ever)— Always beware of building in the following:

- **Emotional content**—Anger, hostility, frustration, hurt feelings, resentment, unhappiness—they have no place in a business meeting, and they don't belong in a memo. If you write an e-mail when you're angry, save it and review it later before sending. You'll be surprised at how even a sentence or two can convey a state of mind that you're better off controlling.
- **Criticism**—Including personal, negative commentary in a written message is always a bad idea. This is true even if you're writing to a subordinate and have good reason to complain. Criticizing someone is a delicate human interaction that should be done face to face. If you must write, stay neutral and perhaps ask questions. ("I didn't see the report on my desk this morning. What's the story?")
- **Humor**—Jokes, quips, and funny references to other people can all be taken wrong—or right. Most of all, avoid irony and sarcasm. These tones don't carry in written form: They demand the visual and aural cues of live interaction. Trying to use sarcasm or irony in written communication can land you in a lot of trouble.

TRUTH 10

Knowing your inside story is the key

The brand new cub reporter came to work in the morning and the editor asked, "Where's the story on the church performance last night?"

"Couldn't do it," replied the young reporter.

"Why not?" asked the editor.

"Didn't see it."

"Why? Didn't you go?"

"Yes, but they didn't give the performance," said the reporter.

"Why in the world not?"

"The church burned down."

Ask, "What will it take to achieve my goal?" But think like Goldilocks—your content needs to be just right: not too much, not too little.

The point of this old journalism anecdote, for our current purpose: *Know what the story is and what matters.* You can't and shouldn't write without substance. Writing is not "a way with words" or, ever, a sleight of hand to conceal lack of substance or thought. We're not saying there isn't a ton of writing like that in the corporate world—only that's it's bad writing.

Figuring out what to say in your message is key. How to do it? Once you've determined goal and audience, you have a good head start.

Let's work backward this time with the simple example we talked about in the discussion of tone in Truth 9. Marian's goal was to get the target people to the meeting and ensure its success. If you were writing the memos, you'd ask yourself, what will that take? What must I communicate to achieve my goal?

For Alpha, the boss: Remind her that the subject is of interest to her so that the meeting is relevant...make the timing convenient, even flexible, because she is juggling so many demands...build in follow-up.

For Mark, the subordinate: He's on the spot. You need him to come to the meeting totally prepared so that he reflects well on you and shows himself off well. In fact, you'd better check out his work in advance. Build in follow-up.

For Joan, your colleague: Let her know about the meeting in which she has an indirect stake (for instance, because her department needs Section A done before it can proceed with its own work). But you don't want any surprises, so the two of you should be on the same wavelength before the meeting itself. Friendliness and goodwill are important. Build in follow-up that reflects this and further strengthens your alliance.

These are simple examples of content mapping. Do the three memos cover the territory? Do you see any missed angles?

Actually, we can point out at least one missing content piece. In the memo to Mark, it isn't necessary to paint the scene as a make-or-break opportunity for him, but including the fact that your own boss will be there to see him perform more than gets that point across. That fact, and a rundown of the other people who will be present, should probably be included in all three memos, as should location (where the meeting will take place) and timing (how long it's expected to last).

Mapping longer projects—Mapping content for a longer or more challenging piece of writing—whether it's a report, an article, a proposal, or a customer contact letter—is just the same. Simply jot down everything you can think of that will help you achieve your goal, and will work with the target audience. What points must you make to state your case? What will show the recipient that he or she should give you what you want?

Don't worry about how the points connect to each other or how to sequence them at this thinking-through stage. Trying to create a formal outline can paralyze many people.

For a complicated writing project, your list can include numerous points of information: facts, statistics, ideas, anecdotes, illustrations, references, and so on. Figuring out what content will accomplish your goals often shows you what additional information to gather.

We're not recommending information overload—the object is always to provide what will achieve your goal in the shortest, simplest way.

TRUTH

Forget outlines—organize your thinking

In school, your teachers probably made you create a careful outline of your essays and papers before writing. Here's the problem: If you haven't thought through your content and made the decisions suggested in the previous chapters, you'll find it horrendously hard to organize the material.

But once you've thought about goal, audience, and mapping your content, organizing your piece logically becomes a snap.

Suppose you're writing a memo to introduce your staff to a new system for reporting their work hours.

- **Goal**—To minimize grumbling and objections to a new system and have staff accept the new way of doing things with reasonable cheerfulness, as well as understand the basics of what they'll have to do.
- **Audience**—One you know well (your staff), and their responses probably run the gamut from those comfortable with change to active resistors.
- **Tone**—"Friendly formal" seems best.
- **Content mapping**—What do you need to say to this audience to get what you want?
  1. Explain the need for a new system briefly.
  2. Announce the start date.
  3. Show how it works in practical terms.
  4. Mention any advantages the new system might hold for the staff.
  5. Make sure they understand that they have no choice.
  6. Offer a way for them to get answers to any questions they may have.
  7. Mention any negative results if they don't follow the new system.

Identify your goal and audience and map the content you need; information gaps will become obvious. So will the best way to organize what you want to say.

You know what needs to be communicated. So all you have to do is organize it—or maybe just check to see whether you're satisfied with the organization that's fallen into place.

You can make choices about what order to put things in, but it wouldn't be logical to start with #3—which such memos often do—because readers won't know why they should care about a new system if they don't know that they must use it.

What's your best opening? A clear statement announcing the new system and starting date. What's a good ending? The offer to entertain questions lets you end on a positive note. The middle offers some choice—#3 and #4 could be interchanged. You can move #5, #6, and #7 around, but introducing a threat earlier may get some readers' backs up.

Try your hand at writing the memo before looking at our version.

Dear Staff Members:

As of September 1, we'll all be using a new system for reporting our time. Here's how it works.

At the beginning of each week, you'll get an e-mailed form (etc., as clearly and briefly as possible).

The new system will give you your paychecks and expense reimbursements two or three days faster because Payroll can process records more quickly when the information is clear.

We'll meet briefly to go over the process on Monday at 10 a.m. in the conference room. Ask any questions about it then, and follow up with me personally if necessary. It's important to get this right because your check won't be issued if you don't follow through properly.

Thanks for handling this change efficiently.—Jim

Note that we added a few things to the plan when we came to the actual writing: announcing the meeting, and the "thanks" at the end. Why? Because neither idea had occurred to us before. As we wrote the simple memo, the meeting seemed like a good way to go in the circumstances outlined, and the closing carries both conviction that the new rules will be followed and a touch of appreciation for accepting procedural change (which, face it, nobody likes).

When the "technical" part of an e-mail needs to be long or complicated, it might better be delivered as an attachment; or you can make it the final section of the e-mail and head it clearly (How to Use the System). Otherwise, readers may get lost in the details and miss important parts of your message.

Writing helps you think

Here's something else that is important about writing effectively when you give it a chance: *What you write reveals the quality of your thinking.* When you really understand what you're writing about, you can describe it clearly. If Jim describes the new system but is hazy about how it works, that will show up in his memo.

To reverse the idea, consider that the reason so much bad writing exists is probably because there's so much bad thinking.

This is all too true in today's business world. Fuzzy, confused writing results from fuzzy, confused thinking. Overinflated claims for a person, product, or organization backed by little or no evidence—what we call "empty rhetoric"—lacks substance and fools no one.

In fact, if you want to be sure you understand a new product, service, or technical procedure, try writing about it (or teaching it to someone else). When you hit gaps in your knowledge or thinking, take the time to remedy them. You can do research, ask a colleague, think some more, review your notes: whatever it takes. Most writers we know have one eye on their writing, the other on a search engine as they work so that they can look up terms and references they don't understand.

You can't fake good writing because it's built on good thinking. But take the trouble to try to write well, and the process will lead you back to your thinking. Improve your thinking, and you'll write better—a win-win situation.

Many professionals—from scientists to artists to business leaders—use *writing as a way of crystallizing their thinking.* Attempting to articulate ideas helps you figure out what you know and what you don't know, and points up what you need to find out.

With relatively short documents, you can choose to just write it all down organically and see what you've got. If you followed the first four steps of our strategy, your material will probably be reasonably organized. Then you can take a closer look and start shifting around sentences, paragraphs, and sections to improve the logical flow of your ideas, facts, and arguments.

TRUTH 12

How to organize is a personal choice

While mapping the content leads you to a natural organization for relatively short and simple kinds of writing—memos, letters, and such—longer and more complicated materials make organizing more of a challenge.

Here are some approaches professional writers use to organize their work when it's complex. These approaches can be adapted for proposals, reports, articles, and other writing projects where you have a substantial amount of information to pull into shape. Some approaches give you ways to build in the organization early on, saving you lots of time later:

There's no one "right way" to organize complex material. Figure out which system suits your style and personality.

1. Divide and conquer—List your major project components that are likely to make up sections in the final document, and give each a separate piece of "paper" (a file) on your computer. For example, if you're assigned to assemble a company history, you might have sections such as Founding and Early History, Important People, Product Development, Current Financial Picture, and so on. Then, as you're assembling information, add the relevant information to the appropriate page or file. Thanks to the computer, you can shuffle the pieces around after collecting them, too. People used to do this on index cards—and some still do—so try that if you're a tactile learner.

2. Create a master list—Skim your pile of raws notes and make a list of the most important ideas or elements. Review the list and see what order seems most logical and rearrange as necessary. Then for each idea or element, find the backup information in your material and add it in directly under the right heading.

3. Color code—Print out your mass of material and go through it with colored markers, matching up the color with a section—green for financial information, for example. You can also mark a piece for great quotes, possible leads, endings, and more so that they're easy to find later. You can color code on your computer screen, too, but because it's hard to do a lot of reading on screen, many people prefer to work with printouts.

4. Bubble it—Some of us learned this method in school. It works best with pen and paper. Put each major idea on paper with a circle around it, leaving plenty of room among the circles. Then draw lines between the circles of those ideas that connect, or logically follow, the other. This gets messy, but you can see the whole complicated picture on one sheet. If you're a visual thinker, it might be the technique for you. The Visual Thesaurus (www.visualthesaurus.com) is an example of this technique. Thinkmap (www.thinkmap.com) describes it as creating "word maps that blossom with meanings and branch to related words."

5. Don't look now—Without referring to your massive pile of notes, and preferably after you haven't looked through it for a day or so, think about what comes to your mind as important (or interesting) about the information you've collected. Many of us find our subconscious has been hard at work figuring this out while we weren't actively focused on the project. Write down the points that occur to you, and then review your material for backup in each category.

6. Tell somebody—Suppose you're reporting on a research project. Ask yourself, what would I say if X (my boss, girlfriend, cube-mate, whoever) asked me why I did this, why does it matter, what did I find? Your answer gives you the kernel of the report and a structure that works.

7. Let your computer do more of the work—There is a lot of project management software available, ranging from open source desktop and Web-based programs to proprietary programs for desktops and the Web. They vary in difficulty and in the learning curve required to use them. Many are collaborative, which means you can work with others who are not sitting next to you but are continents away.

Try some of these systems out and see if they help. If not, don't use them—they are not ends in themselves. Everyone works differently, and there's never only one way to get where you want. Thanks to modern technology, the tools for organizing, reorganizing, shifting material, and making big changes are easy to use.

We won't talk in detail about editing yet, because we have to get the words down first. But here's a liberating thought: You're never—

well, hardly ever—stuck with your initial version. In fact, writers label that version "the first draft," then review it to see what's not working, and fix it.

That means you can experiment with your first draft and not get hung up on details and missing pieces. Will you have to schedule time for editing a major writing project? Yes, but ultimately, you can get the work done with surprising efficiency.

TRUTH 13

Every message you send has a psychological impact

Which of the following e-mails start well? Which do not? Which versions would you keep reading, and how you would feel about each?

#1 Subject: Share a good idea, win a prize

Colleagues: Do you have an idea about how the company can do something better? Improve a system? Reach more customers? Tell our story more effectively?

CEO Jack Martin wants to know, so on March 18...

#2 Subject: Introducing a new suggestion system

Colleagues: Our CEO, Jack Martin, has asked this department to develop and implement a new system designed to encourage staff suggestions, so I'm writing to tell you that on March 18...

Assuming that you prefer alternative #1, notice how much energy it picks up because the writer thought through the goal (to generate good usable ideas) and the audience (busy, cynically inclined employees) and then connected the two with a prize. If you try to imagine the reasoning behind #2, it would probably be something like, "I'm required to make this announcement but expect no results."

Remember that with written messages, all the cues from tone of voice, facial expression, and body language are missing. So it's critical to frame your message by anticipating response.

#1 Subject: The McAllen account

Dear Joe: As you know, at this time of the year, we review responsibilities for individual members of the team, considering general workload and record of accomplishment, and have decided to ask you to take on an additional responsibility...

#2 Subject: New assignment for you—the McAllen account

Dear Joe: Great news! We've decided you're ready to take on the McAllen account, which as you know, is a very important one for the company. Your achievement record has done the trick. I don't think it should unbalance your workload, but if it proves to be time-consuming, we can talk about getting help for you on another account...

Which e-mail would you rather get? The information is nearly the same, but with version #1, you'd be groaning by the end of the first sentence. But version #2 would make you feel good about the new assignment, and valued. Further, from the writer's viewpoint, putting a positive spin on the message leads to supporting thoughts—for example, that help will be provided if needed.

Here's something to remember: *Everything you write, even a garden-variety e-mail, has a psychological impact.* It reflects relationships and sets the tone for how things are perceived. Ten years ago, no one would have written an e-mail message like "Dear Joe"; the information would have been delivered in person, or perhaps in a carefully written memo. But today, the ability to communicate instantly by e-mail has made us very careless about how the way we say things affects other people.

What's the real message?—Note that when you read a message, the physical cues—tone of voice, expression, body language—are missing. So, your innocent e-mail might send an entirely different message than what you meant. That's why you've got to be aware of your tone and how your message may be received on the other end.

If you want to motivate others, or impress those you report to, take the potential of your written words seriously. You'll have a powerful leadership tool in your hands.

When the news is bad—Here's an example of how to think through your choices in crafting a bad-news message.

#1 Subject: No holiday party

Dear Staff: Please be advised that the holiday party scheduled for December 18 is canceled. The reason is that our cost-cutting program requires it.

We hope the company is in a position to reinstitute this tradition next year. Thanks for understanding.

#2 Subject: The holiday party is canceled

Dear Staff: This is news we regret having to share with you. At a time when we are working to trim costs in the interest of maintaining all staff positions, the Executive Committee has decided to forgo the traditional holiday party planned for December 18.

Instead, everyone is invited to gather for coffee and cake in the Johnson Auditorium at 10 a.m. on the 18th—and to bring a toy as a donation to the Homeless Children's Holiday Drive. We can wish each other all the season's joys and also share our good fortune with those who have little.

Which would you rather get? The news is essentially the same, but recipients of the first message might well grumble at this particular cost slashing—we can hear them mutter about executive perks being the last benefit to go—and employee insecurity is being fed as well. Is my job safe? If management doesn't care about the holiday season, will they care about cutting my job?

Message #2, on the other hand, puts everyone in the same boat: We're all disappointed—but it's more important to maintain jobs; who could deny that? It's good to see someone taking responsibility for the decision. And the substitute invitation results from the kind of brainstorming that should accompany situations of this type. It costs little, offers something in the holiday spirit to look forward to, and contributing toys for needy kids reminds everyone of their relative good fortune. It also demonstrates that the company has a heart and it's in the right place, which is very reassuring.

Guidelines for delivering bad news

- Tell the total truth (or as much as you can and is appropriate).
- Show a human face (in a restrained way).
- Acknowledge that someone, or at least a department or unit, made the decision rather than using the obscuring passive ("a mistake has been made...").
- Think about how your audience will react, on every level you can think of.
- Map the content in response to anticipated reaction.
- When possible, come up with some mitigating factor.
- Review the message you craft carefully for how it comes across.

It's not a bad idea to test run bad-news messages by a few trusted colleagues.

These rules don't apply to more serious bad-news situations such as firing people, taking them to task, or transferring them, because you should never use e-mail for what should be one-on-one conversations.

TRUTH

Effective messages lead with strength

Did you freeze whenever your high school or college teachers told you to develop "a strong thesis statement" for your "essays"?

Journalists have a much more encouraging way of talking about an opening. They call it the lead. For a newspaper or magazine article—and for broadcast journalism as well—the opening statement has a lot of work to do: It must pull the reader in, represent the full content of the document, establish accurate expectations, create the tone, and more.

A lead for an advertising or promotional document works even harder to attract attention and set up the reader to view the rest of the piece favorably. In both journalism and advertising, the lead must answer that essential "what's in it for me?" question: Why should I care?

Business communication is not very different. Whether you're writing a letter, memo, report to your boss or a colleague, white paper, proposal, news release, home page of a Web site, or an article for your company's newsletter, the lead must focus your audience's attention and crystallize the core of your message.

The lead can be a sentence, a paragraph, or more, depending on the nature and length of the document. If your message is delivered via e-mail, you should consider the subject line as an important part of your lead.

Professional writers probably spend half their time developing the right lead. That's because when you start right, the rest will follow (although it often works in reverse—many writers create or rewrite their lead after the rest of the story is finished).

Today's business audiences are impatient, wary victims of overflowing in-boxes. When you make readers work to know what your message is about, or to understand why they should care, you lose them.

Here's a useful way to think about the lead. It must:

- Tell your specific audience how your subject relates to them.
- Indicate why the subject is important in general.
- Suggest what you will be asking them to do.

Someone we know who was trained to write in the army says he learned: "Put the bottom line on top." It's a good rule to follow. To see why, just look critically at the e-mail messages you receive in the course of a day. How much time do you spend figuring out why each person wrote to you, if the message is of interest, and whether you should read, file, or delete it?

Wouldn't it be a more perfect (and efficient) world if you knew exactly what every message was about and what interest it held for you, after reading just the subject line and the first sentence or two?

Realistically, it's the same for every kind of writing, from letters selling insurance to reports to your boss, a note to the Purchasing Dept. and announcements of all kinds. If the first paragraph or so doesn't catch us, we stop reading. Suppose you receive the following memo about a training seminar.

Subject: Training seminar November 8

I'm pleased to tell you that my department has been charged with planning and implementing a series of workshops to upgrade new managers' skills. The resources were provided by HR after an analysis of staff capabilities and company needs. Sessions will be offered every month for the rest of the year. On November 8...

So far, who cares? If you're having a busy day, will you be drawn to read further and find out what HR is congratulating itself for? Here's an alternative.

Subject: Nov. 8—Learn to be a great presenter

You're invited to a special morning workshop on November 8th: "How to Deliver Dynamite Presentations." This major leadership skill was pinpointed by HR as key to manager success, and we're flying an expert in from Chicago to lead the session.

Maybe you have something better to do on November 8, but you understand the opportunity, right? You'll read the rest for details if the workshop appeals to you. If it doesn't, you recognize your disinterest faster.

The point is, when you're writing, lead with strength: Start with the best reasons why the people you're addressing should be interested. They'll always want to know how the matter affects them and will read down to the details if it does.

Provided you've done the brainstorming work we recommend, you now just need to look at your content mapping list. Pick what's most important, relevant to your audience, and intriguing if possible, and make that your lead, in whole or part.

When you're writing the memo, how do you arrive at a lead like the second one? Your preliminary thinking tells you. You already know:

- **Your goal—**To get good attendance at the workshop, which is voluntary.
- **Your audience—**New and middle managers who badly need to develop presentation skills but may not want to take the time, may fear speaking before an audience, or doubt the session will be helpful.
- **Your content—**This workshop can directly help the target audience polish a major skill...the skill matters to the company... the session will be taught by a very good person...plan to reserve the time. Plus details about where and when, how to sign up, and so forth.
- **Best way of organizing the material—**For the medium you're using—e-mail—you must quickly get your points across so that the recipients don't filter the message out.

Now you can build a lead paragraph that covers as many of the content points as you can.

TRUTH 15

To succeed, cover your ground and remember "the ask"

With most writing situations, the middle section is the nitty-gritty, descriptive information. It tells the reader how to do something, gives details for an event, specifies the items to be covered in an assignment, provides technical specifications, spells out the reasoning behind a decision, or whatever else is necessary.

Your lead will have set up the reader by defining the subject, setting the tone, and establishing a "what's in it for me?" appeal. It might be only a sentence long, as in many e-mails, or it could be several paragraphs of a long document.

Whatever the medium, the middle needs to follow through and cover your content list in the logical order you figured out earlier. It's just a matter of getting the content down on "paper."

For example, if you're making an assignment to someone who reports to you, the middle of the message would detail what the assignment consists of. You could do that as a bulleted list or as a narrative, making sure your instructions are clear if the person is new on the job or hasn't done this type of work before.

So, Mike, I'm asking you to analyze the new spare parts program for our widget shipments. We need you to determine whether the program is working properly, whether the parts are arriving on schedule, and whether they are in good condition when they arrive. Please send me a report detailing your findings in each of these areas. If you think it would be helpful to include a graph or chart to show your results visually, please append it. If you need one of the accountants to work with you, feel free to use my name when you ask for help.

The lead paragraph would have stated the assignment and deadline, and any context needed to do the job well. The closing paragraph might repeat the deadline and any contact information Mike might need.

The middle of most messages contains the technical content: how to do something, event details, descriptions or specifications, the reasoning behind a decision, backup for your position.

If you're writing a response to a customer complaint—for example, a letter complaining that a mail-order dress didn't fit—you'd use the opening paragraph to say something positive (such as, "We are delighted that you've been a customer of X brands for three years, and that you chose Style Y, which we just introduced this season."). Then you'd use the middle of your message to explain nicely why the error may have happened, because, for example, this style fits differently.

Write the middle quickly—Our recommendation is to write the middle quickly: Get it all down and then go back and edit. That way you can review for the right amount of information: what's not needed, what's missing. You can improve clarity and conciseness; find better words and linkages where the expression is awkward or confusing; and, of course, fix mistakes.

As far as substance goes, you've already done all that prep work in thinking about goals, audience, and content, and part of the payoff now is that it's relatively easy to write the middle.

Depending on the length and nature of your document, you can employ various kinds of graphic devices to help organize your informational material, break it up, and make it accessible: subheads, bold lead-ins, bullets, numbers, and more.

Every message should end well—Many times, the conclusion needs to bring home the action you're requesting. In sales terms, it's "the ask." And remarkably, sales trainers find that in many cases, salespeople fail to ask for the business. Let's consider how the messages presented as examples in preceding sections should end.

The training seminar invitation

Please mark your calendar and e-mail me back by October 30 so we can reserve a seat for you.

The "share a good idea" memo

March 1 is the deadline to receive your suggestions—send as many entries as you want. We'll read them all and present the best on the 18th, crediting the contributor, along with awarding the prize. Good luck!

Don't flub "the ask": Salespeople often forget to close the sale with a direct request, and so do business writers.

The "here's a new account for you" e-mail

Congratulations, Joe, I know you'll do a great job. Give me a call this week and we'll talk about any questions you may have.

The assignment given to Mike

The best person in Accounting to talk to is Meg White, who's familiar with the project. And of course, call me with any questions or problems.

I look forward to having the report in hand no later than April 10.

The customer complaint letter (in this case, you need to come to a resolution)

I'm sorry you were disappointed with how the dress fit, and we are happy to issue a full refund on receipt. In future issues of the catalog, we will take care to clarify how this particular model is sized. Thank you again for being our customer.

The ending gives you a chance to reinforce what you're trying to get across—an apology and commitment to solving the problem in the case of the customer complaint, for example, or a motivational feeling in the case of Mike's assignment. At the same time, you should state or restate the need for a response if appropriate.

The conclusion closes the circle: The message has been set up to engage the audience (lead), deliver the needed substance (middle), and clearly conclude (end).

Better to ask twice—A word of caution: If "the ask" is a request for the recipient to take an action, such as to "e-mail me back by October 30 so we can reserve a seat for you," it's usually better not to hold this information for last.

Putting "the bottom line on top" makes more sense in these days when everyone is too rushed to read the whole memo. The e-mail recipient may never get to the end, so anything vitally important belongs in the lead. Use the ending sentence or paragraph to reinforce the thought. For example, you can start with "Reserve a seat by October 30 for...," and your close can be "I look forward to hearing from you by October 30."

TRUTH 16

Your goal and audience determine the best way to communicate

Jessica had worked for Kate, her new supervisor, for six months and found her boss's style cold. So she wrote a long e-mail explaining how Kate could get better results from the staff by praising them more and not giving preferential treatment to a few.

Was that a good idea? Obviously not. Even if the boss wanted Jessica's advice—not all that likely—a personal meeting would have been a better choice. The same is true the other way around: Jessica would much prefer an in-person critique of her performance to a written one.

Here are some guidelines to help you choose the best communication channel. Remember to take the individual person into account, especially his or her age. There's a real generational divide on preferred communication modes. *When you're addressing multigenerational audiences, use multiple channels.*

E-mails, letters, and even texting play important roles, but so do phone calls and meetings. When personal response and interaction really matter, don't write—find a way to meet.

E-mails—E-mail is useful for sending brief notes to colleagues, such as telling them about meetings and checking their schedules; inviting people to lunch; giving a quick progress report; providing or requesting information; giving updates after a meeting and making assignments; and more.

Did you notice the common thread? E-mails are best when they're short, to the point, written with an immediate purpose, and also, geared toward people who use and like e-mail.

What about those times when you need to deliver a lot of information? Many of us work virtually now, so almost everything we write is sent by e-mail, and in most organizations, sending reports and other materials by interoffice mail doesn't cut it.

The principle still holds: Keep e-mails short and deliver bulky material as attachments. Then the e-mail serves as a cover letter.

Never forget that e-mails tend to live forever, thanks to backups, e-mail storage, and archiving. Whatever you write in an e-mail can come back to haunt you, in legal, professional, or personal ways. If you don't want to see your message on the front page of the newspaper, don't e-mail it.

Letters—There are occasions when a letter, whether sent by postal or e-mail, is right. Among these are when you need to thank someone; request an appointment, favor, contribution, or interview; introduce yourself to a potential employer; provide a reference or recommend a colleague; or congratulate someone.

Letters work to promote your business in the community or via a network and to tell clients about a new service or special offer. You need letters to cover a proposal, grant application, or résumé submission; and to create a formal or official record, such as a complaint, legal announcement, performance review, resignation, or offer of employment. And you need letters if you want to present your views to your elected representatives or a newspaper.

The common thread? Formality. Sometimes, also, your communication may need to be a bit more formal because the recipient is higher up in the company or older than you.

Letters used to be necessary for signatures, but many organizations now accept faxed or even e-mailed (typed) signatures as valid, although you should check to be sure they're legal for your purpose.

Paradoxically, a letter can communicate more warmth and individuality than an e-mail. This makes letter writing a good choice when your goal is relationship building. Letters have their own downsides, including the fact that copies can be made and circulated, or the letters themselves can be saved and archived.

Telephone calls—Sometimes it's better to pick up the phone. This applies if the other person responds better to the sound of the human voice and whenever it's important for you to hear a personal reaction. The telephone also works if you're untangling a knotty issue that's best talked out, need an immediate response, or want the conversational interaction you don't get with written communication. A call can be followed up with an e-mail or a letter giving details or reinforcing what was said on the phone.

Texting—For business purposes texting, is increasingly widespread, because it's so convenient for our on-the-go, on-the-road work styles. Often it's the only form of communication that's practical, given differences in location, time, and accessibility. Texting's downside is clear: The messages may be cryptic, hard to decipher, or both. It's also a generational thing, in that many people beyond their twenties have never texted and may not want to.

Meetings—Face-to-face communication has many advantages, mainly because body language, facial expressions, and tone are clear to everyone. Although corporations use video-teleconferencing and even sites like Second Life for group conversations, nothing really gets the message across as well as sitting across a desk or a table.

So, when should you aim for a personal get-together? When the topic is sensitive or possibly hurtful; you're delivering criticism or bad news; a brainstorming session is needed; you're making a sales pitch to a potential major client; a major decision or organizational change is being made; cultural differences could affect communication; or when a team culture needs to be forged.

For example, a software manager in Boston, supervising a work group in Bulgaria, may find that e-mail, conference calls, and videoconferences are difficult because of language-use problems, time differences, and the like. Getting on a plane and sitting down face to face at least once may resolve long-standing differences.

Unified communications—Another way of sending a message is called "presence," or "unified communications." This means sending a message—by e-mail or voice—that's delivered to the recipient via cell phone, desk telephone, desktop computer, laptop, fax, BlackBerry, or similar device. This lets you communicate with recipients wherever they happen to be, in the fastest possible way for each.

The bottom line—There's no one answer to which communication medium to use. Among the variables, consider the age and tech savviness of recipients; the time/speed required; the potential language, cultural, and physical barriers; and the nature of your message.

TRUTH 17

The best writers don't write; they rewrite

"A writer isn't someone who writes, but someone who rewrites." This is a basic principle of professional business writers, and if you ask one what it applies to, he or she will say, "Everything." The reason is simple: The first version is usually awful.

But editing is probably not a subject you were taught during your school years, unless you were lucky enough to have had an amazing English teacher or personal mentor. Like most people, you were probably left to figure out how to improve your writing on your own.

You may have found this hard to do. However, some guidelines and practice can make all the difference. It's also important to recognize that you're not trying to imitate twentieth-century academic models. Your goal is to produce clear, effective writing that connects with your readers in the twenty-first-century digital world. This doesn't require laboring over rules of grammar and structure. In fact, you might find that what you read here conflicts with some of the things your teachers told you.

Real power in writing happens when you view your document—including an e-mail—as a first draft, to be cut, tightened, sharpened, fixed.

What kind of writing do the guidelines apply to? Every kind. In fact, they're even more important to newer forms of communications like e-mail. E-media demand brevity and instant readability. You must know what you want to say and say it clearly and concisely.

Here's the first guideline: *Think of everything you write, even an e-mail, as a first draft with errors you can fix once it's "down."* Every professional writer writes, then edits. Don't see this as demanding more time, but as freedom to write more spontaneously rather than getting hung up on word choice or structure. Plan to edit. This enables you to take advantage of the other ideas that follow.

Think short words—Words with one or two syllables are best. Use words with three or more syllables sparingly and deliberately, and when you can't think of a shorter substitute. Why? First, the fewer syllables your words have, the faster your writing can be read, and speed is everything. Remember your readers? Short of time, impatient, jaded?

> Short is in: The digital age values short words, sentences, paragraphs and documents that we can read with speed and instantly understand.

Furthermore, short words are always clear and understood by nearly all English speakers. They feel natural—because it's the way we talk. And they feel more trustworthy than complicated words. It probably relates to our language's history: The original Anglo-Saxon supplied the short, basic words, whereas long ones from Latin and French were grafted on later. This means we have many more word choices than most languages, but for contemporary business writing, choose the short ones where you can.

Short Words Work Best

Use	Rather Than	Use	Rather Than
wordy	verbose	also, too	additionally
use	utilize	basic	fundamental
fake	fraudulent	many	numerous
start	initiate	common	prevalent
show	demonstrate	often	frequently
help, aid	assist or facilitate	try	attempt
rule	regulation	about	approximately
need	requirement		

Think short sentences—Long sentences slow down reading, and you lose your audience. Who has time to figure out what somebody is trying to say?

Various studies on preferred sentence length indicate that averaged over the whole document, 12 to 15 words is about right. The sentence you're reading now contains 23 words, but averaged with the four preceding sentences, the number of words per sentence is 14.

You don't have to accept this as a hard and fast rule, but notice that "ideal" length is likely a lot shorter than you thought. If long sentences plague your writing—and trust us, you are not alone—you can shorten them in three basic ways: Cut unnecessary words, cut unnecessary thoughts, and break long sentences into short ones.

Think short paragraphs—Research on readability—how fast people read and how much they understand—shows that a paragraph should average 45 words, and no more than 65. That's only three to five sentences.

You can prove this to yourself by flipping through a few magazines of different types. To read pages of dense type with few white spaces, you've got to be *very* motivated. Not that much of what you write is likely to be a "must-read," even if you're the boss, so give your readers a break—literally. Use short paragraphs, subheads, and visuals (if called for) such as charts or graphs.

The cure for long paragraphs is easy: Break them up. Avoid chopping in the middle of a thought, but three to five sentences usually cover an idea. You may need to watch out for your transitions, though—more on that soon.

Do the guidelines apply to e-media?—Yes! Even more than with "traditional media," e-mails, Web sites, blogs, letters, proposals, and everything else need to be reader friendly. You need shorter words, sentences, and paragraphs. Particularly on screen, people don't like to read a lot of material, and they don't like to scroll.

Think short documents—What does nearly every business audience have in common? A short attention span. Think about your own impatience: You want to know the point of an e-mail or letter before you read it. You want to know the gist of the message immediately. You don't want to figure out what's important—you want the writer to do that for you.

In general: Half the length can double your chances of getting a document read.

TRUTH 18

Rhythm and transitions make writing move

Base your writing on straightforward, simple sentences. Teachers used to call these "declarative" sentences. For example, "John broke the stick." "Amanda is going to the overseas conference." "I want to tell you a story." "Mr. Macklowe needs the report tomorrow." "Most immigrants learn to speak English in two years."

But, you say, if I use only sentences like that, won't my writing get boring and choppy? Yes. You've only to look at a textbook to see something like this:

John Black was a Pilgrim. The Pilgrims suffered from oppression in England. Black emigrated to America in 1654. He bought land in Massachusetts. He built a large house. He was elected town governor in 1666.

How can you avoid this mind-numbing effect? Easy...

Alternate the simple sentences with longer (compound) sentences that have two or three sections, or clauses. It's just as simple to write:

John Black was a Pilgrim. Because Pilgrims in England suffered from oppression, he emigrated to America in 1654. He bought land in Massachusetts and built a large house. In 1666, Black was elected town governor.

Simple, straightforward sentences give you a base but make them flow by attending to rhythm and using easy structural variations.

Still not riveting, of course, because the content is dull. But by varying the sentence length and fooling with the wording a little, you at least have a paragraph that flows reasonably well.

Here's a work-a-day example:

Hi Mike:

Thanks for inviting me to the meeting on Friday. It definitely sounds interesting. I'd like to learn about Socratic selling. However, I'm starting my vacation on Thursday. I'm sorry that I can't be there. Please keep me on the distribution list. I'd like to come next time.

Versus

Hi Mike:

Thanks for inviting me to the meeting on Friday. It sounds interesting, especially because I've always been curious about Socratic selling. However, I'm starting my vacation on Thursday. I'll have to miss this meeting, but please keep me on the distribution list so I can come next time.

Build in transition words to connect facts and thoughts, so your writing will seem logical, persuasive, and inevitable.

Again, some changes were needed to adapt the sentences to the new structure. What's been accomplished?

This technique gives you rhythm—try the "reading-it-aloud test" with both versions—and it pulls the reader along. Moreover, hasn't the message become more convincing in expressing appreciation and regret? The stilted version sounds like the writer is just going through the motions of responding.

Pay attention to transitions

First, check to make sure every sentence connects to the next one. If it doesn't, try to alter the wording or the order of ideas so that it does.

For example:

We've found four problems with the new software. It was purchased from Dann Associates, and we got a break on the price. The backup system doesn't work...

Better

We got a price break on the new software from Dann Assoc., but we've found four problems. First, the backup system doesn't work...

In the first example, a reader is thrown from one point to another—from "problems" to where it was bought/price break and then back to "problems." In the second version, the connections are clear—notice that the word "problems" at the end of the first sentence in the second version leads right into the list of problems.

It's all about relationships. Each sentence should follow logically from what precedes it, and the part that relates to the following sentence should be put as close to it as possible.

You may need transitional words to make connections clear—we used *but* in the preceding example.

Here are some other transitional words: *and, while, although, however, nevertheless, because, moreover, instead, alternatively, otherwise, further, additionally, sometimes, similarly, again, as well as, meanwhile, specifically, for example.*

Phrases can also provide transitions: on the other hand, in spite of, to the contrary, in line with, not only, and so on.

Using transitions to suggest sequence is often helpful. Ideas can be numbered, for example; you can introduce them as "first," "second," and so on. You can profitably make the logic of even a short e-mail clear with sequencing techniques and an introductory sentence:

We see four problems with the new software:

First...

Fourth and last...

This lets people know how much is ahead of them and gives them a sense of satisfaction as they progress through the sequence—especially as the end approaches. Good speakers use this technique all the time to keep us listening to them.

Pay close attention to the transitions between paragraphs, too. Each transition needs to link logically to the next one. Gaps force readers to figure out the connection you didn't make. Wherever the paragraph ends is where the following one should pick up. Transitional words are useful here, just as they are between sentences.

You don't need to know what a document is about to see how the following phrases might connect a paragraph to the one preceding it or following it:

In that case, however, how did they end up in trouble?

With passage of this regulation, the industry climate changed.

Here's the background.

But this is not a strategy for success.

Here's another way to see this.

In spite of this experience...

I see three ways to accomplish the goal.

In conclusion...

TRUTH

19

Less can be a whole lot more

Cut descriptive words—adjectives and adverbs—to a bare minimum. Mark Twain wrote in a letter to a 12-year-old, "If you find an adjective, kill it." He may have been objecting to the flowery, ornate language that was common in the nineteenth century; his own prose is simple and eloquent. Today, heavy use of adjectives and adverbs doesn't suit our cynical modern tempo, and—just as important in the business world—it works against believability.

What do you think if a car salesman tells you this?

This is the most amazing car in the world, and its features are incredibly advanced.

What has he said? Nothing. He's used abstract descriptive words to shortcut substance. It's silly for him to think he's succeeding, if the goal is for you to buy the car; but why do so many organizations communicate in similar ways? Depending on adjectives and adverbs may be the source of a lot of empty corporate rhetoric. This tendency shows up especially in press releases, which often tout "our world-acclaimed, industry-leading technology company's groundbreaking new gadget."

> If you cut empty rhetoric, unnecessary words, thoughts, and needless repetition from your writing, you will double your impact.

The solution again is simple. When you review what you've written, cut out all those words unless you are certain that they're absolutely needed. Try to *show* rather than *tell or describe*. Stick to the facts and honest ideas. If you don't have any, get some or don't communicate! That's what today's readers want.

If you believe that your boss or the company's executives want those three-dollar words in your organization's news releases, Web site pages, or newsletters, make an effort to explain the logic in showing, rather than telling. Give them examples. They may agree once they see the point.

Your goal: Strip it. Here are three ways:

1. **Cut words and thoughts that don't contribute to your message**—You've already defined what you want to accomplish, and this tells you instantly what's essential and what's not. Look for extra ideas that don't reinforce your argument or lead the reader in a different direction, and throw them out. And look for words that interfere with your message because they're not needed, not totally appropriate, or not clear.

   This is from a major organization's news release (names changed to protect identities):

   The Web site has a number of features including an interactive map of the site plan that is complete with descriptions of the numerous features. There are also RSS feeds that allow subscribers to get notified when there is an event or any news. In addition, the Web site also offers its users the ability to support the Shipley project by writing to County or Town elected officials.

   Try rewriting this before you read our edited version at the end of this section.

2. **Use an eagle eye to spot repetitions in words, phrases, and ideas alike**—For example:

   Because the necessary supplies came late and were not delivered on schedule, unfortunately, Project J, I am sorry to say, will probably be delayed.

   Better:

   Because the supplies came late, Project J will be delayed.

   In addition to tightening the language, you can often substitute better words or phrasing and improve readability. For example:

   If in your opinion you think John is ready to handle a major project like Project J, let me know and I'll consider assigning the project to him.

   Better:

   If you think John is ready to handle a major assignment, I'll consider asking him to handle Project J.

3. **Look for repetitive sounds in your copy**—The read-it-aloud technique presented in Truth 2 is a wonderful help in self-editing. The idea: Reading what you've written aloud will immediately show you where you need to make changes. Doing this highlights

unnecessary repetition and awkward constructions so you know exactly what to cut or reword. Here's a sentence we wrote for an early version of this chapter:

This method will immediately identify the need for clarity.

Try to say it aloud and all the *y* sounds force you into a sing-songy rhythm. So we rewrote it this way:

This method will instantly show you why it's important to be clear.

Remember, these are *ideas* for you to write with, rather than *rules*. Absorb the ideas, and you won't have to worry about mastering hundreds of rules. You'll know how to figure out how to improve your own writing. (See a problem with that last sentence? Two *how to's* in a row? Try rewriting it.)

Rewrite challenge

There are many ways to edit—that's why it's not a science. Here's one way the paragraph cited in the first editing tip can be rewritten:

The Web site features an interactive map of the site plan that lists all its features. It also has RSS feeds, notifying subscribers of news and events, and links to county and town officials so Shipley Project supporters can contact them directly.

TRUTH 20

Passive thinking and jargon undermine clarity

Kill the passive and commit to action—you've probably seen or heard of this concept before, and it's true: Cut the inactive verbs from your writing wherever you can and you'll liven up your writing enormously. So what's inactive? It's probably easier to show you than to define the term:

The road was crossed by the chicken.

You may think that no one would say this rather than "the chicken crossed the road," but here are some examples of everyday business language that contain the same weakness.

Passive and inactive writing come from lazy thinking and often a futile desire to avoid responsibility. Have the guts to say "I."

Example 1

The report on the company's credit situation was written by James Coopersmith. (passive)

James Coopersmith was the author of the report on the company's credit situation. (inactive)

James Coopersmith wrote the report on the company's credit situation. (active)

Example 2

The accounting department's annual audit was supervised by a committee appointed by the board of directors. (passive)

A committee appointed by the board of directors supervised the accounting department's annual report. (active)

Example 3

To have edited your work after having written it may turn out to be a good thing. (inactive and convoluted)

Editing your work is probably a good thing. (active)

Or, better:

Make a practice of editing your work after writing it. (active)

Example 4

Upon review, you'll find that mistakes have been made. (passive)

When you review your work, you'll find that you've made mistakes. (active)

Why do we tend to write in a passive way so often? Sometimes because we're lazy, but often because we're pussyfooting around responsibility: We don't want to say outright that we're talking about *I*, *we*, *you*, and so on. The last passive sentence in the preceding group is an example of the "divine passive"—nobody did it, it just happened ("mistakes have been made")—and politicians, bureaucrats, accountants, and many others often present their failures that way.

But if you try for simple declarative sentences, you can avoid this temptation. Most writing will benefit from using *I* and *you* much more often. People want to know who did what and why they should care. Remember "What's in it for me?"

A major reason why you shouldn't use passive constructions is that they dilute and weaken your meaning: "John was accused by Tom of being an idiot" is a lot weaker than "Tom accused John of being an idiot."

And relying on *is, are, was, to be*, and related "state of being" words drains your writing of fire. Often an active verb can substitute for a long wordy phrase. "The new processing system is a replica of the one used by Johnson" is weaker than "The new processing system replicates Johnson's."

"The maple trees are very colorful in autumn" pales before "The maple trees blaze their autumn red and gold."

It's the difference between describing and showing. Good writers put endless effort into replacing passive and inactive verbs with active, interesting ones that provoke graphic images. Those active verbs get there in the editing process—rarely in the first draft.

So for real power, improve your writing by working in action words with some life to them. Or should we say, rather, to spark your writing, pepper it with words that sing.

And murder that jargon—Exclude or explain everything that could be misunderstood by anybody else—jargon, acronyms, abbreviations.

Even highly educated people complain that their peers often use words and phrases that don't have clear meaning: industry jargon that masks a lack of thought...acronyms that demand research to track down and sometimes cannot be found...abbreviations that hinder comprehension, at best. Just try to say exactly what you mean. Here's what can happen when you don't:

We'll provide an overview here of the key enhancements in the new driver that impacts both XYZ Server 8500 and LMN Server 9310

ABCD-based data access, as well as provide a glimpse into the future roadmap for our ABCD support and commitment to enterprise-class interoperability within LMN Servers.

One problem with using jargon is that we forget that it's jargon and that its meaning may not actually be clear. Here are just a few examples from the business world:

synergies	gain traction	110%
scaling up	leapfrog	shovel-ready
value-added	leverage	360-degree thinking
functionality	mission-critical	

What are the jargon words and terms in your business or profession? Are some understood only within your company?

Jargon vs. specialized knowledge—If you insist that other specialists in your field will know what you mean, note that there's a difference between assuming a common base of knowledge and depending on jargon. If you're a scientist writing to other scientists, for example, you can count on their understanding scientific ideas without having to start from the ground up. When you use scientific terminology, you needn't define it—but if someone did check a dictionary, the meaning would be totally precise.

This isn't true of jargon, which typically consists of industry buzzwords that don't necessarily have a shared meaning. "Turnkey training" is an everyday example—educators use it to denote "train the trainer," while in business, it can mean ready-to-use training programs, or training that produces ready-to-go workers, or training that is customized to specific goals, and more.

It is also very risky to use acronyms that some readers won't understand and to import abbreviations from a medium like text messaging into another, even e-mail. Why? Maybe you can assume many texters understand the same protocols you do. But what about someone you're e-mailing who doesn't use text messaging?

Consider the woman who picked up some of her 13-year-old daughter's expressions when she started using a BlackBerry device, like writing "lol" to denote *laugh out loud*. She thought she was being very cool until a prospective client she was courting—a middle-aged CEO—asked her sheepishly, "Why are you sending me lots of love?"

TRUTH 21

You don't need grammar drills to spot your writing problems

When your writing is weak and awkward, look for clues that signal a problem. You can almost always simplify, shorten, and clarify. Sometimes you can solve a tough problem by throwing out the whole sentence or writing a new one. Be aware of the following signs that a sentence can work better, and you're on the road to better writing without having to think about grammar.

Heavy use of the word *of*

The CEO of the company that produces cosmetics made of pomegranate oil is of the opinion that demand will grow.

Why not:

The CEO believes that demand for the company's pomegranate oil products will grow.

What's wrong with this?

Few things affect quality of life as much as the removal of the waste products of our civilization.

Better:

Few things affect quality of life more than removing our civilization's waste products.

Too many *to's*

Jones needed to do something to revitalize the community relations project.

Better:

Jones needed to revitalize the community relations project.

Too many *-ing's*

The lab is focused on developing nano-engineered particles that can be much more powerful in catalyzing combustion.

Better:

The lab focuses on developing nano-engineered particles that can more powerfully catalyze combustion.

Too many *-ion's*

The dimension of the problem is an indication of the infiltration of bad work habits.

Better:

The dimension of the problem shows that bad habits are infiltrating the workforce.

Try to use only one *to*, *of*, *-ing*, and *-ion* per sentence.

When you follow up these clues, you'll find they'll lead you to fix many of the same problem areas we've identified in other ways—using the passive tense, writing awkward or confusing constructions, including unnecessary words, and so on.

At this point, you may want to ask: Will following these guidelines give me choppy, dull writing that makes me look like a simpleton? *Au contraire*. Consider the Gettysburg Address, *The Wall Street Journal*, Emily Dickinson's poems, Ernest Hemingway's prose, and the Declaration of Independence. All communicate the most complex human thoughts in simple, direct language that reaches people on a deep emotional level.

Editing tip #1—Distance yourself from your work by a day or two when you can, or a few hours, at least. It'll be far easier to spot anything awkward, ambiguous, or unnecessary, and easier to find better ways to express your meaning. This definitely applies to important e-mails, and just about everything else.

Editing tip #2—Find a writing buddy. Beyond finding mistakes, a backup reader notices whether any "attitude" is showing—hostility, for example, when you write to someone you don't like. Your writing buddy can be a friend, colleague, or even someone at home.

Editing tip #3—Sharpen your own editing skills with this activity: Select something you wrote that's at least a page long and then cut it by half. Edit out what's not necessary or important and anything off-message. Reword for clarity, simplicity, and brevity. Work in transitions as needed. Simplify sentences and tighten everything you can until you've reduced the piece to 50 percent of its original size.

Now compare the two versions. Which is better?

Try to expand your edited version by 25 percent, adding back in some of what you cut.

If you edit well, you'll discover that it's harder to add anything back in than it was to cut it out in the first place. Lesson? Editing, even by such an arbitrary standard, improves writing.

Use your computer's Readability Index

Your computer gives you a tool, the Flesch Index, to instantly check how readable your writing is. It tells you the reading level of your material, the percentage of passive sentences, the average length of words and paragraphs, and the number of sentences per paragraph.

To bring it up, you may first have to be sure there's a check mark next to "Show Readability Statistics," which is usually under Tools/Spelling and Grammar/Options.

> Editing your own work is easy when you know what clues to look for and how your computer can help.

Then, with a document on the screen, click the "Tools" menu, and then click "Spelling and Grammar." Run the spelling check, and a Readability Statistic box will appear.

The first set of figures gives you an instant way to check how well you're following the guidelines in this book and using short words, sentences, and paragraphs. The second set gives you some major clues about overall readability. *The higher the percentage of passive sentences, the harder it is for readers to relate to the material.* Passive sentences tend to be formal, abstract, and static. Aim for between 5 percent and 10 percent.

The Flesch Reading Ease Score shows how many people will understand the document. A score of 100 indicates one that is so simple that "everyone" will understand it, and 0 denotes a very complex document: A score of 50 to 60 is a good goal, but when your audience includes numerous people with little education, aim higher.

The Flesch-Kincaid Grade Level indicates the years of education required to comprehend a document. *The Wall Street Journal* aims for eighth-grade level, which is hard to achieve. It's easier to write with more complexity.

Using the index—After completing a first draft, bring up the Readability Index. If the Reading Ease Score and Grade Level show that the writing is hard to understand, review your document to cut down on passive sentences and shorten them, replace long words with short ones, and break up long paragraphs. Recheck the index. It's a great way to improve your writing.

TRUTH 22

Use e-mail to communicate in the fast lane—powerfully

Did you hear the one about the company CEO whose indiscreet e-mail about the company president was forwarded to the board of directors by a disgruntled executive? Or about a giant firm's financial managers who naively thought that "delete" actually erased their e-mails about illegal accounting practices?

Such high-profile anecdotes pale in comparison to the thousands—maybe millions—of e-mails that are just poorly conceived and written and, as a result, damage countless careers and relationships. And, most people don't even realize when it happens.

Electronic communication is how we interact with most people now. Your e-mails may be the only thing your colleagues, bosses, clients, and customers ever see from you. Your professionalism, or lack of it, is your calling card.

All the principles of good writing that we've been talking about apply to e-mails. When is a carelessly thought-out e-mail with indifferent spelling, grammar, and punctuation okay? Never. E-mails are kicked upstairs, sideways, out of the loop into new networks endlessly. Their ghosts may live forever in company storage servers and archives, backed up for disaster recovery, government regulations, and legal reasons. They can and do come back to haunt you.

It's also a mistake to think that you have two or more "selves"—one who writes "good" and proper e-mails to the boss, for example, and a second self who writes casual, sloppy e-mails to people you're not worried about impressing. Often our friends and business associates travel in the same circles. When you behave unprofessionally—and poorly done e-mails are definitely unprofessional—you risk losing opportunities you'll never hear about.

The bottom line: Supervisors take note of well-written e-mails, especially over the long run, and so do clients and colleagues. And every day, e-mail gives you at least an equal opportunity to screw things up, maybe irretrievably.

All the principles of good writing apply to e-mail. Knowing this puts you way ahead of everyone who's contributing to the flood of careless, badly thought-out messages.

We haven't even mentioned the efficiency factor: Poorly written e-mails that deliver information badly, cause confusion, and waste time are so common that hardly anyone complains about them. One carelessly done e-mail about a meeting time, sent to 20 people, could easily cost each of them 15 minutes worth of e-mailing and phone calls to resolve. That's five hours for just a simple example.

So write good e-mails. Here's how.

The challenge: You want to go to an expensive conference run by your industry's association and need your boss's sign-off. You know the budget is tight.

Your goal is clear. *Your audience* is your boss—someone you probably know quite well. Or do you? If you've never done a "formal" analysis of your supervisor, invest the time, especially if you feel as if you're not on the same wavelength. Use the criteria in Truth 7 for audience analysis and don't skip the "what's in it for me?" criterion. Consider personality types, too, for clues about what works with different kinds of people.

Remember that your memo may be channeled up the managerial line, and to lateral departments, so remember that your audience often goes beyond your immediate target.

Tone: Go for respectful and business formal, without groveling.

Content: What might make your case? Business advantages all the way—probably your boss doesn't want to think you'll be enjoying the surf in Hawaii. Specify:

- What you'll learn and what that will do for your employer
- Whom you'll meet
- The value of having your company represented
- Any supporting information (such as, you haven't been to a conference for two years, or the last one you attended produced a long-term client)

Don't forget to cover:

- How your work will be handled in your absence
- That your trip will cause your boss absolutely no inconvenience

How to organize? Start with the order of your content list and see if it works.

How to lead? For an e-mail, the lead is the subject line, plus the opening sentence or two of the message. Here, typically, you clearly identify your subject and give the reader a positive view of your goal.

The working rule is: *Put the bottom line on top.*

One way to write this e-mail

Subject: Major Industry Conference Opportunity in June

Dear Elaine:

I request your approval to attend the Bottomline Building Association conference in Cleveland June 11-13.

Three special presentations this year directly relate to major company initiatives: project financing, working constructively with local government, and downtown revitalization. Several of the country's top experts will speak, and I'd plan to follow up with personal meetings. I intend to bring back problem-solving ideas and guidelines.

Additionally, I'll have opportunities to meet on a collegial level with some of the 1,500 attendees from around the world, so I'll have chances to scout the competition, find leads to collaborators, and even clients. Companies including Marvel Construction and Worldwide Fabrication will be there in full force.

I plan to be a very active representative of this company and work to raise our international profile.

This will be the first major conference I've attended in two years. As you'll recall, my participation at a regional-level BBA conference then resulted in a long-term relationship with CGA Development.

Workflow should present no problems—I've completed the Smith project, and Jerry is fully prepared to back me up. And of course I'll keep in close touch so I can handle anything unexpected.

The conference information is attached. I will appreciate your approval, and any input for representing the firm.

Review your e-mails—every one

Now, evaluate: Did you make a good case? Leave anything out? Is it right for the person you're addressing?

And, the most important thing: Would you approve the request if the relationship were reversed? Put yourself in the boss's shoes and see whether what you've written would persuade you to send the writer to the conference. If not, rewrite!

TRUTH

23

Good subject lines say, "Open sesame"

Simply put, the purpose of a subject line on an e-mail is to get the recipient to open and read the message. Like a good lead, the subject line should attract the reader and get him or her involved in the message.

In a business setting, it's best to stay away from anything that can be construed as "cutesy," suggestive, bold, or nasty. The trick is to come up with subject lines that tell the story directly, clearly, and with as few words as possible. Here are some examples, good and bad, taken from the authors' mailboxes:

Don't get trashed: Come up with subject lines that tell the story directly, clearly, and concisely.

- **Winners**

  Case Study PDF on Server Ready to Proof

  Chapter 11 Needs Two More Examples

  Speaking Opportunity Offered to Your Club

  More Information for ASTD Conference Attendees

  IT Dept. Progress Report, June 30

  Referred You to Potential Client, Collegiate Institute

  IABC Nov.18 Meeting Location Changed

- **Second Place**

  Shipping Dept. Is Low on Manila Boxes

  Three Editing Questions on Article

  Writing Tips for Communicators

  Info for Members

  Add Items to Agenda

  Sept. 24 Query

  Meeting Notes

  Need a Reference

  Need Immediate Answer

- **Losers**

  Reply #2

  Yet Another Question

  Advertorial Inquiry

  Get Me Off This List!

  Web Site

  News from Ralph Jackson

  Started a New Biz

  Good Day

  Need Your Help

  Don't Miss This Opportunity!!!

In considering these subject lines, you may notice a few commonalities:

1. The winners tend to be strong, concise, direct, and self-explanatory.
2. Those in second place need additional information for the recipient to get a good grasp of what the e-mail contains. In each case, the recipient is left saying, "What does that mean?"
3. And the losers are in that boat because they say too little or too much, forcing the recipient to make a judgment call on whether the message is spam or a useful bit of information. If, for example, you don't know who Ralph Jackson is, you probably will press Delete quickly.

Thinking through subject lines—Because e-mails are forever, taking the time to make yours work is time well spent. With subject lines, this means making sure the subject is clearly identified and will pull the reader into the body of the e-mail. The urge to press Delete is strong, so don't give your recipients a chance to do it. (Of course, if the e-mail is in fact irrelevant to recipients, a clear subject line won't help you.) If what you're communicating is important or even urgent, get that across right up front but be specific: "Important matter" doesn't work; "Important meeting for accounting staff" does.

There are additional reasons to take trouble with subject lines:

- **Finding specific e-mails—**Good subject lines make it easy for people to find e-mails again when necessary, which can be quite often. Don't be surprised if someone actually thanks you for this someday—how often do you feel frustrated trying to retrieve an e-mail from a co-worker who customarily labels them all "Greetings"?
- **Change the subject line of the thread if the topic changes—** A "thread" is the sequence of e-mails on one topic that may loop back and forth between sender and recipient several times. If the topic changes in, say, the fourth round, change the subject line. It can be maddening to search a thread labeled "2010 Projections" for a reference to new-construction costs that you know is in there somewhere.
- **Topics and legalities—**In business, e-mail subject lines should serve as specific headings for the topics they contain. There are even legal reasons. Both subject lines and content are used during the discovery phase of court cases to find and identify companies' or executives' unethical or illegal behaviors. If your subject line can be misconstrued, there may be unintended consequences.

And one more thought: If you have an e-mail address that's not professional sounding, change it. Receiving a business e-mail from "Dis-a-kitty" or from "imbozo32" will not exactly strike the recipient as a business communication. Your e-mail may even wind up in the company's spam filter before it reaches the intended recipient. Although corporate, educational, and scientific e-mail addresses use a business-type format, some independent business owners may be more free spirited—often to their own detriment.

Most professional-looking is to use an address with a domain name, which is easy to obtain even if you don't have a Web site.

TRUTH 24

Know your e-mail do's and don'ts

We've become a nation of scanners. We screen every incoming message for relevance and importance, and if we decide to read it, usually give it a rapid review and stop as soon as we feel we've gotten the gist. Beyond the fact that there are so many demands on our attention, we don't like reading a lot of text on screen, and there are physical reasons for this—it tires our eyes.

Take this into account in crafting your e-mails. Here are some things to keep in mind.

E-mail may be your communication lifeline, but it has many limitations: Messages work best when they're short, sweet, and oriented to a single clear purpose.

E-mail do's

- Put the bottom line on top. Don't make people guess why you're writing or what you want.
- Take the time to write strong subject lines that work as leads and clearly identify the subject—which shouldn't need saying again but it does. People will delete anything frivolous, irrelevant, and unclear.
- Make e-mails short and stick to one subject. People read quickly and distractedly—if you ask two questions in the same e-mail, often you'll get a response to only one. Buried points and subtleties will also be overlooked.
- When you have a major goal to accomplish, you might plan a series of e-mails rather than trying to jam a lot of information into one. If you need to spell out to a group what's involved in various stages of a project, for example, consider covering one area at a time.
- Use attachments when your subject is necessarily long and complex, unless you know the person or company won't open it. In that case, separate the body of the message graphically from the "cover letter" aspect by using a headline or bold lead-in.
- Think short and concise in every possible way: words, sentences, paragraphs.
- Aim to avoid making the reader scroll. People don't like to and won't.

- Make sure your text size is readable on screen. It should never be smaller than 10 point; 12 point is better. Also ensure that your line length doesn't exceed 60 characters, or it may run off the end of the recipient's screen.
- Organize clearly and simply. Bullet points and numbered paragraphs are useful for people who might not read to the end of the e-mail. When you use numbers, start with a sentence such as "Here are three items that need your attention." You can also put this information into the subject line, such as, "Two Questions re Finch Contract."
- Follow up. Don't rely 100 percent on cyberspace. When an e-mail is important to you, check that it was received. Rarely will people mind. It's risky to assume that they got your message, given the increasing number of filters that might identify your message as spam or high risk.

E-mail don'ts

- Don't forget to ask for what you want at the end, even if you have to repeat.
- Don't use fancy graphics that require HTML; many people don't have the option or won't use it, and your message may look terrible. Some corporate spam filters also reject such messages.
- Don't use a color other than black for your typeface, a fancy typeface, or a background with a design on it for business e-mail. Not only does it look unprofessional, but it also makes reading the message more difficult and these elements may not display properly on the recipient's screen.
- Don't use all capitals, which make it look as if you're shouting and make the message hard to read, too. And avoid using all italics, which are also hard to read.
- Don't use e-mails for subtle messages—and keep away from sarcasm, irony, and for the most part, humor. It is easy to take such things the wrong way when voice and visual clues are absent.
- Never, ever e-mail anything you don't want to see on your boss's wall or on anyone else's computer anywhere in the world.

- Don't let emotion, anger, or criticism rear their heads in e-mails. Breaking up, resigning, or firing someone via e-mail is *very bad form*.
- Never click Send without proofreading, checking to be sure the e-mail is addressed to the person you want to send it to, and reviewing your attachments. If you often forget to attach documents you intend to, writing a reminder sentence into the e-mail is helpful, such as, "Report 5A attached." It helps you remember to attach it and also helps the recipient, in case he or she fails to notice the attachment itself.

TRUTH 25

Writing good progress reports is worth your time

Let's look at something that most people hate doing but can really make you look good if done well: regular reports to your boss on your work progress or what you've been doing all week.

Goal—Think past the idea that your goal is simply to get an onerous task out of the way. You want to demonstrate that you've used your time well, made progress on your current project, and get the support you need (supplies, equipment, resources, discussion).

Audience—The "what's in it for me?" idea is always important. So why does the boss impose this task on you and your colleagues? Distrust? No...he or she hopes to hear you are performing well. In a department with more than a handful of staff members, a supervisor simply can't observe what everyone is doing and track the big picture. And he or she can't hold enough meetings to interact one on one and drag any problems or questions to the surface.

Always think through *why* you're asked to write something: This tells you the logic to follow and enables you to do a good job on assignments that are routine but important.

Ultimately, too, your boss needs to report on collective progress to someone higher up the ladder. Your reports may also be essential for billing clients.

Tone—One more thing that just about everyone you write to has in common—feeling strapped for time. Almost every e-mail you write will be read by an impatient and distracted person. This affects your presentation approach: You'll want to be businesslike, efficient, and tight.

Content—What does your goal and audience analysis already tell you to cover in your progress report, regardless of the specifics of your job?

- A rundown of how you used your time (a proportional approach is probably fine)
- If your work is project-oriented, what progress you've made, and whether you're on schedule

- Any problems you're encountering
- Any help you need to do your project, or job, well

Organize—The content map is logical; try using that order.

Begin with a strong lead—In an e-mail, that's the subject line, plus the first sentence or so of the message. Some subjects call for a catchy subject line. That's when you're competing for attention. This is not one of those occasions. Your boss will not respond favorably to "What Willie Did the Week of March 14th." You're better off with "Progress Report: Week of March 14."

The middle and end will flow naturally because you've got your message planned, so let's start with the message lead and see where it takes us.

Dear Sam:

Here's my progress report for the week of March 14.

General Allocation of Time:

50% on the Wise-Allen project.

25% closing out the details on the Fineman project.

10% participating in meetings and industry networking.

10% training and supervising the interns.

5% working on the database reorganization.

Personal-Time Activities:

After-hours social meeting with Rod Blaine of SatSun, a good prospect for our services.

Spoke on international commerce careers at the high school Tuesday night.

Major Project Progress:

Wise-Allen is on target in most respects: The supplies have been ordered, the staff is briefed, and the working plan is three-quarters developed. Obtaining additional staff support is running late, however.

Fineman should be wrapped up within the 10-day framework.

New Contacts:

Met and talked at some length with Brad Savitch, VP of Manson Inc., at the JVNC meeting on Wednesday, and plan a follow-up call to request a meeting. His firm may be interested in our international support service.

Intern Program:

The interns are doing productive work and having a good experience helping with the database work. I met with each of them for 15 minutes to check things out.

Problems: *As noted above, there have been delays in getting the help we need from HR. We need time from specialists for the technical aspects of Wise-Allen—I'll attach a list—which could cause serious delays in completing this stage of the project. Would you consider placing a call to HR to let them know this is important, or drop them a note? Thanks. —Bill*

Now we have a draft of the e-mail. Next...

Evaluate—What comes across overall—what would be your impression of the writer? If the message doesn't meet the stated goals, or doesn't seem right for the audience, you'd adapt or change it at this point. Notice we added in a "Personal Activities" section because in the course of drafting the memo, it seemed like a good opportunity to remind Sam that this employee goes above and beyond the standard work-hour framework and expectations. But whether you should do this depends on the personality of your recipient.

Of course, you may work in an office that demands far more detail or provides a form to fill out, especially if billable hours must be documented. Adapt the ideas to your own environment and the people you work with and for. Even company guidelines can be applied with imagination. Take your activity reports seriously, and you'll see opportunities to give a strong impression of your capabilities and advance your long-range goals.

TRUTH

Letters: They live! And you need them

Even in the digital age, businesspeople use letters for building client relations, proposals, references, invitations, and all the other formal occasions of the business world. In each case, the step-by-step process will focus your thinking so you come up with the right presentation.

Of course, the line between an e-mail and a letter can be pretty blurry. We're not necessarily defining a letter as a hard-copy document that is snail-mailed. It might be e-mailed, sent as a whole message or as an attachment. *Regardless of the delivery system, you need to consciously decide when you are writing a letter and think it through as a letter for it to achieve your goal.*

Letters are not dead artifacts: They nurture relationships and are the way to go when you're asking for something, marking a formal occasion, introducing yourself, and more.

"Can we meet?"—Let's turn to a classic business situation: writing a letter to request an appointment to present your product or service to a prospective client. Suppose you spoke with someone briefly at a networking event. Here's one way to follow up:

Dear Jenny:

I enjoyed talking with you at the AALG meeting yesterday. You mentioned your interest in touring the Amazon, so I'm attaching information about an ecolodge a good friend of mine stayed in and highly recommends.

And, I will very much appreciate an opportunity to tell you about my company's services as an HR outsourcing firm. We have served a number of companies in the medical lab industry, and were successful in saving them considerable money while improving their customer service systems.

May I have 10 minutes of your time to show you what Outsource Strategies can do for you?

Where did the ecolodge bit come from? Surprisingly often, you'll see a chance to follow up even a brief exchange with some information of interest: the name of a book or restaurant...a source of data or new software...or best of all, a connection to someone the other person would like to know.

"Thanks for your time"—Margaret applied for a job recently that was just high enough on the corporate pecking order for her to make a brief appearance at a board of directors meeting. She knew she was on a short list but that there were at least two other candidates. So she went home and wrote a letter to every member of the board.

Briefly and respectfully, she thanked them for their time and expressed enthusiasm for the opportunity. Then she summarized in one carefully written paragraph why she was convinced she could do the job outstandingly. She took trouble to make the letters look good and delivered them to each board member's workplace.

As the only candidate who wrote letters, Margaret got the job. Of course she was qualified, but so were her competitors. Writing the letter, and writing it well, gave her the edge, or so she is convinced.

It's always good to take the trouble to thank people for their time, preferably by letter. This applies whether it involves a prospective client, a reference, a referral, a favor, a subordinate's extra effort, or a supplier's blowing out all the stops to deliver on time.

"Thanks but no thanks"—Often, the letter format is the most appropriate when delivering unwelcome news is the task. This is where looking at the situation from the other person's viewpoint really pays off. Whether you're discontinuing a supplier, rejecting a job candidate or canceling an office perk, your goals will generally be:

- Deliver the information unequivocally, leaving no room for argument or misunderstanding
- Minimize hostile reaction to your organization—and to you personally
- Reassure the recipient in some way

These goals suggest that you need to take responsibility for the decision—no cop-outs like "it has been decided." You should show caring (but not emotion), and when possible, cite a reason for the action. If you can legitimately hold out hope for the future do so, or include other encouraging ideas. For example, you might offer the unsuccessful job candidate a connection to someone you know, or tell someone who has been denied promotion how to qualify next time.

Cushioning a blow this way, with real substance, works much better than the verbal cushioning often recommended for framing bad messages. The latter is sometimes called the "sandwich" approach: You start by saying something as positive as possible, then deliver the bad news, and close with expressions of good will and/or other reassurance. For example:

Dear Jim:

For more than 10 years, it's been my pleasure to work with you and your company as a supplier of Part 32B. Unfailingly, we have found that you delivered on time, as promised, and met our specifications.

Right now, however, I must tell you that next year's contract has been awarded to another firm. A-Plus has grown so much that a national distribution network has become imperative.

Should circumstances change or other needs arise, be assured I would look forward to working with you again.

That's okay, but consider that the opening paragraph delivers an unwelcome note of suspense ("uh, oh, I hear an ax falling...") and moreover, doesn't suit today's accelerating tempo. We think most recipients will prefer the direct, get-it-over-with approach:

Dear Jim:

I am sorry to let you know that next year's contract for Part 32B has been awarded to another firm. The reason is that Ajax's growth the past year has made it imperative for us to have a national distribution network.

Please know that your excellent track record with us remains very much appreciated, and I hope we'll have opportunities to work together in the future.

Of course, there are times when it's appropriate to sugarcoat the facts. If you're writing to a customer, for example, responding to a complaint or denying a request, you'd want to begin by setting a positive context and a feeling of relationship.

Generally, it's a good rule to keep negative messages brief. Just as political history shows us that cover-ups always create much worse fallout than the original transgressions, bad-news messages that meander can generate bigger problems than the news itself. Generally, the less said the better.

Letters build relationships

Writing a letter of introduction**—One of us recently got a snail-mail letter from a doctor who'd taken over a retired doctor's practice and wanted to keep his predecessor's patients. He began:**

I would like to take this opportunity to introduce myself. I am a board certified dermatologist practicing in this area since 2002. I trained at...

And on with more than a page of credentials and a list of dermatology diseases he's treated that, well, made our skin crawl. The writing is technically fine but fails utterly. It's the message that doesn't work—why not?

The doctor appears to have been writing to impress his peers, not prospective patients. Had he considered his audience's concerns, he might have come up with a content map that includes: I'm a warm, comfortable guy who cares about his patients...my office will be run efficiently and respectfully... and I'm a very good doctor. The writer doesn't actually need to say he's a warm person, but he needs to demonstrate it in his writing tone. For example:

When people can't see or hear you, they decide who you are based on what you write and how you write it. With many letters, coming across as a human being is an important part of your content.

Dear Ms. White:

As you know, Dr. Andrew Grant retired in June, and I am taking over his practice. I don't know you yet personally but want to introduce myself.

In addition to bringing strong experience and credentials to Hartley Street Medical, I plan to provide a comfortable, responsive environment...

Notice that once we take the trouble to produce a good lead, the rest readily falls into place. The basic information can be categorized (for example, training, experience, local connections). It is perfectly okay to use subheads or bold lead-ins in a letter, and/or bullets. A good business letter does not necessarily require a narrative flow all the way through.

Of course, a personal salutation and real signature help personalize a letter.

A resignation letter—Suppose you've got a new job and it's time to resign from your old one. Do yourself a favor—don't use e-mail for this! Usually it's best to tell your immediate superior in a private meeting and follow up with a formal letter. It will circulate through company channels and be filed forever, remember. So formal business style is the way to go.

What's your goal? Don't make it revenge: It's human to think about getting even for all the small snubs, resentments, and relationship issues that develop over time. But that's the fantasy—you're the grownup and have to do what's best for you in the long run.

Your goals in 90 percent of leave-taking situations should be...

- To ensure you get good references every time you consider a new opportunity.
- To maintain cordial relations with people who remain part of the same industry.
- To leave people feeling good about you—and themselves.
- To leave a foot in the door. You may want to come back someday on a higher level or work for someone who will move on and be in a hiring position elsewhere.

How to do this? Consider it from the other people's perspective. Unless you've been messing up and they're happy to see you go, they may feel that the company investment in your training was wasted and may also feel let down personally, even abandoned. So you need to express appreciation for the opportunities afforded you, regret at leaving, and positive things about your experience at the company and with the particular person. It's a good idea to remind everyone that you will remain part of their industry and that they may benefit from putting a good face on your leaving.

Dear Jack:

This is to officially confirm that I will be leaving Martin Brothers effective Friday, May 30.

I will miss many things about the company, which has given me wonderful opportunities to learn and grow over the past five years. Because of this, I am ready to take on a challenging opportunity in the accounting department of ABA Inc.

Continuing in the underwriting business gives me great satisfaction, and I look forward to seeing you and the rest of the executive team at events and professional meetings. It would be terrific if chances to work together develop.

I've enjoyed working with you more than I can say. I wish you the best of luck in all your own future endeavors.

What if the work experience and relationships were not so hot and writing such a letter makes you feel like a hypocrite? Before sending a more honest message, remember that leaving on a high note can mend many fences retroactively. One of us had a colleague whose work came under criticism at top levels. Seeing the writing on the wall, Jenn quietly applied for other jobs and landed a good one. She then wrote gracious letters to several key people about how sorry she was to leave and why the new opportunity was irresistible. Result: They were delighted to save face and behaved with equal graciousness. Jenn always gets glowing references from the organization—and a smile when her name comes up.

What if you're fired or laid off?—The same principle applies—doubly. Look to the future and put a good face on your leave-taking. Don't come across as angry, pitiful, or vindictive. Maintain your dignity and you'll reap future rewards, not the least of which is self-respect.

Handwritten notes for great networking—A consultant we know makes a point of knowing what's happening in the lives of her clients and, whenever she finds a reason, sends handwritten messages on notepaper. For example:

Dear Jerry, I was absolutely delighted to learn that you are this year's recipient of the Tumblewood Award. The association could not have made a better choice. Your contributions to our industry have been outstanding for many years, and I look forward to applauding them at the awards dinner in June.

Dear Joan: A little bird told me that your son has been chosen as valedictorian of his graduating class. What a terrific achievement—I know how proud you must be. I just had to let you know how happy I am for your family.

Think this is tacky? The writer is always thanked effusively and often finds her notes prominently posted in her clients' offices. Because few people bother to send such messages anymore—especially handwritten ones—a small investment of time makes her stand out. She connects with clients and colleagues in a far more sustaining way.

Cover letters matter, big time

When Simon applied for a new job, the cover letter he submitted with his résumé made the "10 worst letters of the month" list at the company where he wanted to work. The letter was posted on a bulletin board in the HR office, distributed all over the company to people who looked forward to a monthly joke, and was no doubt sent by some recipients to their personal e-mail lists.

Like many job applicants, Simon had taken a lot of trouble to polish his résumé, and then dashed off a quick cover note with little thought. The result was an embarrassment of poor content, language, grammar, spelling, punctuation, and even capitalization.

We think it's fair to say his letter didn't succeed.

When employers ask for cover letters (even with online applications) along with your credentials, they mean it—they want a letter, not a spontaneous-looking e-mail or text message. They will most definitely judge you on how well you've written it.

> Cover letters are often make-or-break factors for job applications, but they're hard to do well. A clear sense of goal and audience tells you what to say and how to say it.

This is perfectly fair. Employers want to see how you handle yourself in writing, which most consider a critical skill. If employed, you may not be asked to draft proposals or articles, but you'll be communicating often by e-mail and probably other written media. In many cases, you'll be representing your employer. The better you can do that, the more of an asset you will be.

So, badly written letters are the first filtering mechanism that consigns your résumés to that circular file, the wastebasket. Those who do the filtering love it when your cover letter fails: They always get more applications than they want to read. And the higher up the company ladder you aim, the higher the writing standards you're expected to meet.

Writing the cover letter—You're answering the following ad for a marketing job, which specifies a cover letter. If marketing is not even close to what you do, even better—practice putting yourself in someone else's shoes.

Position Open: Assistant Marketing Manager

Algorhythm Guitars, America's third largest manufacturer of guitars, is creating a new position in the marketing department due to recent sales growth. Responsibilities include strategic planning support, presentations, working with the sales team, and interfacing with top management. Requires relevant marketing experience, excellent communication and people skills, ability to create and track new programs, and effective team management.

Let's assume you've got some basic qualifications, but it's a stretch. How do you plan your cover letter?

What's your goal?—To get the job? Sure, but narrow that down more exactly and see what happens. Your first goal is to keep your résumé out of the employer's wastebasket. Your next goal is to get the recipient to read your attached résumé with a favorable bias. So you do *not* have to include all your qualifications and personal history. In a way, you're introducing yourself through the cover letter, just enough to make you a person of interest. Your résumé will carry the conversation from there.

This is important: Think of it as divide and conquer. Just about every communication should contain the least information that will accomplish your purpose. How liberating is that?

Who's your audience?—You may be responding to the Human Resources department, the Marketing department head, the company president if the firm is small, an unidentified person's name, or an anonymous e-mail address. Often your application will proceed through a series of checkpoints—when you're lucky.

So, what's the useful way to define the audience? As businesspeople. No matter how trendy the organization, when it hires, it's a business proposition.

What does this tell you about tone?—When you anticipate multiple audiences with different standards and interests or can think of no way to visualize your reader, be formal. It's always appropriate in an arena like the job application. This doesn't mean you should write stiffly or in an old-fashioned way. The piece should be gracefully written and, if possible, project some personality, which résumés can't do.

Content—What points to make? Brainstorm your experience for matches with the ad's requirements. Assuming you are a suitable candidate, your résumés should be proving that you meet the specs. But what will help you rise above the competition? Perhaps...

- You play an instrument yourself.
- You worked two summers in a music store.
- You're a member of Toastmasters.
- You co-managed a successful marketing initiative that involved strategic planning.

How to lead and follow through—Your opening should immediately identify why you're writing, confirm that you meet the basic requirements so you're not wasting their time, and introduce your special strengths as soon as possible. Here's one way:

Dear Ms. Rinehart:

I'm a marketing professional with more than five years of experience in the music industry, and I'm very interested in the new assistant marketing manager position at Algorhythm.

My five years of related work experience are backed by formal training, and I've developed my presentation skills as an active member of Toastmasters. As a serious amateur guitarist, I can bring an in-depth understanding of your company's market to the role. My early work as a music store salesperson gives me an even broader base for creative strategizing.

Recently I conceived and co-directed a successful marketing initiative that involved working with the company's leadership and the sales team. The results have measurably increased this year's sales figures.

I look forward to exploring my qualifications with you in person.

That's all you need.

With cover letters for jobs, remember to echo their specs and buzzwords—our short letter references most of them. This will not be seen as unoriginal, but as understanding what the ad said and establishing yourself as a kindred spirit. So read any ad you answer 10 times to absorb its content—and its tone.

Good reports generate action

Do you sometimes feel buried by requests for reports? You're not alone. With fewer and fewer face-to-face meetings taking place, and so many collaborations to manage, report writing has become a big staple of office life. It's a basic way to keep everyone updated, and therefore needs to be treated as a major business skill. Most organizations require a constant flow of reports: progress and project reports, survey reports, budget reports, committee reports, technical reports, and more.

In general, reports follow some pretty standard patterns: They introduce the subject and explain the reason for writing them; provide "the meat" (what happened, what was learned); in some cases, explain how results were obtained; and usually offer recommendations (where do we go from here?).

In the business world, a successful report is easily defined: It's one that is read, and results in action. The role of good, clear, concise writing is obvious—if the right people don't read your report, you've wasted your time. Write to be read.

Your plan of action—As an example, let's say you're a marketing director and must report on whether your company should venture into emerging international markets. For a high-stakes challenge like this, it helps to put your strategy "on paper" early on to guarantee the right substance—you may discover some missing components that require research. Think step by step:

Goal—To persuade management to market the company's tennis racquets or golf clubs in India, Laos, and China and to open offices in two of the countries. *Writer's sub-agenda:* To position himself for the director's role in the new marketing ventures.

Audience—Chief executive officer, company president, board of directors: a sophisticated group with their eyes right on the bottom line.

Tone—Business formal; objective with understated enthusiasm.

Content—The writer should ask, "What will it take to persuade management to my viewpoint?" Possibilities:

- Overview of the immediate options in specific countries
- Analysis of how selling each product would coordinate with the company's current marketing and production plans

- Financial predictors, such as statistics from a consultant's survey of emerging markets; government forecasts; five-year sales/investment cost projections
- Analysis of each country's political and economic climate
- Balanced recommendations for/against opening the offices

Even within a formal business framework like the report, you make your case best when you deliver writing that moves well and engages the reader.

Organize the material—Starting with the overview is necessary and natural. To make the rest flow logically, you could organize by product—tennis racquets and golf clubs—or by country. Because the basic issue is whether to open offices in each place, organizing by country is probably best. A section for each would cover, first, economic and political climate because strong negative factors here could make the rest irrelevant. Then you'd go on to cover the various predictors of success, fit with company marketing and production, and end with conclusions/recommendations—next steps.

Writing the lead—some options:

- Start with an anecdote that emerged during your research, such as another company's remarkably profitable experience.
- Start with a rhetorical question: "Is it time for ABC Inc. to bring its products to the emerging Asian markets?"
- Start by putting the new ventures into a broader perspective. For example, "ABC Inc. has led the industry for a century by capitalizing on just the right time to make dramatic marketing moves. Now it is time to consider..."

Follow through on writing the middle, based on the organizational plan, section by section. And at the end, summarize your well-reasoned, balanced recommendations and outline the next steps needed to implement your thinking. The end might also include an inspiring statement or a vision of a great future—but don't raise hopes to unreasonable levels.

Write a table of contents for any report that's more than a few pages long.

The executive summary—This is best done after the document itself is finished. You are likely to need executive summaries for business plans and proposals, as well as reports. The executive summary has a bright future because the more time-pressured everyone in business feels, the less people want to read long, complicated documents. Because managers and colleagues may read the executive summary and nothing else, it needs to be self-contained: a piece of writing that can stand on its own if separated from the report itself. Think of it as a call to action. Communicating a sense of urgency may be appropriate.

Start by carefully reviewing the report. Then shape the piece in its own right, aiming for...

- A good title (not just Executive Summary)
- A strong lead, which makes the subject's importance clear and perhaps introduces the recommended action
- A middle section that explains what was done and the main points that support your recommendations
- Conclusions and recommendations

If your report is directed at more than one audience, consider creating different executive summaries for each that take account of their varying "what's in it for me?" viewpoints.

How long should an executive summary be? The classic advice is to make it no more than 10 percent of the document's own length. It can be shorter than that, providing it does the job of defining the subject and its importance, presenting your findings or actions, and offering your conclusions and recommendations.

Review and edit—Always edit your work to sharpen the writing and check for a good level of detail, meaning one that makes the writing interesting and credible, but doesn't interfere with its flow. Statistics, charts, graphs, and other supporting materials can be added in an appendix. *If the report was prepared by a team, it's critical to edit for consistent style, tone, and so on.* As always, you make your best case when you deliver writing that moves well and engages the reader, even within a formal business framework.

TRUTH 30

Organizing complex projects isn't that hard

Elaine, assistant manager of an environmental services company, was handed a challenge: a dense, disorganized, and meandering "white paper" (a report on a major issue) that dealt with an important aspect of the firm's work, remediating polluted lakes and ponds. Her assignment: "Fix it." Where to start?

It's always easiest and most efficient to plan a substantial writing project in advance, before plunging into the research and actual writing. But sometimes a project just grows out of hand. At other times, you've already accumulated masses of information on the subject and, paradoxically, knowing so much can be paralyzing. And sometimes, like Elaine, you're handed a half-baked document and asked to clean it up.

When you're assembling an involved document that presents a lot of information, first make people care. Then sequence the sections to build a natural logic and end with the big picture.

The step-by-step writing method that enables you to plan and follow through on a complicated writing assignment helps in a retrieval effort as well. The process is similar. Here's how it can work—with variations, of course, according to the nature of your subject.

Clarify goals and audience—With a major project, this often *demands* a team approach. If you're dealing with complicated technical information and a morass of possible content that needs to be translated for a general audience, you'll need help from the subject experts. If you are the subject expert, you'll find it helpful to get input from people who communicate effectively. Use team thinking to brainstorm goals in relation to audience.

In the water pollution example, the original focus of the confusing paper was to report on a regional survey the company had done to identify contaminated bodies of water. A think-tank session revealed that even some participants were unclear on basic aspects of the problem, and that many members of the target audience—government administrators, civic leaders, developers, and the general public—would have minimal understanding.

This led to a decision to refocus the paper as a "Water Pollution 101" education piece. The agreed-upon goal was to make people care about the problem, present an analysis of the roadblocks to fixing it, and promote exploration of solutions. The sub-agenda—raising the company profile as a relevant service provider—could be kept unobtrusive so the paper would be perceived as done for the common good.

Content mapping—The articulated goals and better sense of audience made it fairly easy to figure out what content would work. Agreement was quickly reached that the paper should include the following:

- History of why the problem exists
- Clear explanation of why it matters to everyone in the region, and demands attention
- Rundown of current relevant laws
- Analysis of what's wrong with the laws
- Analysis of why little progress has been made toward fixing the problems
- Recommendations toward solutions
- Organization's credentials, to validate its viewpoint
- Examples of successful processes and outcomes

Research and analysis—With a contents hit list in hand, the already written material was reviewed rather easily to see which sections might be used or adapted, and what was missing. The content holes were filled through research and discussion with the subject experts. Keeping the content areas in separate files, and developing each one separately, sidestepped the confusion of having to deal with so much information.

Organization—With all the pieces at hand, it was quickly agreed that it was logical to begin with "why care?"—the "what's in it for me?" concept: If readers didn't see the importance of the subject in the first 10 seconds, they would read no further. From there it was natural to move on to why, given the situation's seriousness, so little progress had been made.

Next, to set the stage for making recommendations, the organization's expertise should be established. The rest of the sequence took shape naturally. When it came to presenting an analysis of each major problem, it seemed most effective to cap each element with the recommended solution. Thus, the final sequence:

1. What defines a polluted body of water, how many there are in the region, and why it matters
2. Reasons why little has been done regionally to clean them up
3. The sponsoring organization's credentials for addressing the problem, briefly
4. History: Where did the problem come from?
5. The relevant laws: each briefly summarized
6. Analysis of what's not working and recommended initiatives toward solutions
7. Conclusion: Recap of what healthy ponds and lakes can mean to the region, how the organization can help, vision for the future
8. Appendix:

    Three case studies of successful programs

    Details on the company's regional survey

    Glossary/definitions

    More detailed backgrounder on the company and its services

In sum, attack big challenges by being clear about who your audience is and what you want to accomplish...break down the elements into specific components...figure out a logic to your argument so you can identify what's missing, as well as how to sequence your material...adjust the organization to make the best use of your material and to make the most powerful argument. Put supplementary material that might slow the flow into an appendix.

These ideas apply directly to proposal writing, reports, and other long documents.

TRUTH 31

Well-crafted proposals win

Two technology companies regularly apply for government defense contracts by responding to RFPs (requests for proposals). One reaps a modest ratio of success. While the executives usually voice a "win some, lose some" attitude, they are always surprised when a bid fails, because "our engineers are the best in the business." The second company lands a steady stream of contracts, some of which are a distinct reach. What's the difference in how the companies bid?

Company #1 leaves the proposal preparation to those who understand the specs and the work. Company #2, while basing the content on its engineers' input, delegates the writing to a small cadre of managers who write skillfully.

Good proposals win; it's that simple. Even with competitive bidding, price can usually be interpreted in terms of extra value and proven ability to deliver. The process for crafting strong proposals also applies to internal proposals—like a service offer to another department or a case for introducing new technology.

Goal and audience—You're competing for a business opportunity. Who will review what you submit? Realistically, a whole series of people with different degrees of knowledge who probably have far too many proposals to plow through. This makes readability valuable. Also, they probably are armed with a checklist of "plus" and "minus" factors, and may review proposals electronically to see that the requisite keywords and elements are included.

So if you have an RFP or a less formal announcement in hand, your first step is to read it 20 times to pick up the keywords and identify the underlying agenda. Look beyond the obvious and analyze: What problem is the company trying to solve? Can you read between the lines to understand what keeps the management up at night? Your content should address that.

If you've met with the clients, visualize them: Reconstruct the conversation and note what mattered to them and how they talked about their needs. Play that back in the proposal, and you may reach them on an emotional level.

Tone and style—Even when an RFP's language is convoluted, your writing must be clear, simple, logical. Strip the modifiers. Keep words, sentences, and paragraphs short. Eradicate promotional hype and claims without evidence, such as, "We're the most technologically sophisticated company in the industry."

A proposal doesn't necessarily have to be a cold, dry document. Often it's good to show a passion for the work you do. People like to deal with those who make it clear that they truly care about their product or service.

Content mapping—What must you include to come out the winner, whether your firm is selling telescopes, psychological services, architectural design, or salami?

- Demonstrate a complete understanding of the business and project and describe what you'll provide in those terms.
- State how you'll do it—time frame, staffing, resources.
- Present your company résumé in tight form, tailored to the project needs.
- Present the advantages you offer, such as experience in the industry, awards, technology infrastructure, special expertise.
- Backgrounds of the principals who'll do the work.
- Comfort factors: evidence that your company can be trusted to do the job right, on time, and within budget—testimonials, track record on similar projects, references.
- Essentials such as fee structure, exclusions, project schedules.

The headline and lead—Unless you're filling out a specific form, you need a headline, perhaps a subhead, and a lead. Generally, headlines can be simple and matter-of-fact:

A WATER SYSTEM PROPOSAL FOR THE TOWN OF BROMPTON, N.Y.
Presented by H2O Products Inc.

Try for a good, benefit-oriented lead. For example, "ARV Media proposes to create a radio campaign that will make Jake's Paint a household name." It's also perfectly okay to use simple, straightforward leads such as, "XJY Architecture is pleased to present this proposal for Barton City's first solar-powered library."

Build your middle—Put the content in logical order in marked sections. Use subheads, bold lead-ins, varying typography.

Conclude—According to what feels right, end with a confident summary statement. You may need an executive summary, a table of contents, and an appendix with backup.

All the principles of good writing apply to proposal writing, which has the clearest bottom-line result of anything you'll ever write. The language may need to be formal and it needs to sound objective, but that doesn't mean it should be stilted and dull: To pull people through a proposal, the writing must be crisp, sharp, active, and convey enthusiasm. Use the "say-ability" test. Use good transitions between sections and paragraphs.

Some more guidelines:

Do...

- Use a team approach and good teaming strategies if different departments are involved.
- Thoroughly research the organization you're pitching and your likely competitors.
- Brainstorm what your firm can offer that others can't or might not think to include.
- Call the company and ask questions (intelligent ones), which may even establish a bit of a relationship.
- Go for evidence of what you can produce or perform.
- Edit a team-produced proposal so sections and styles meld.
- Create the final document with good graphic principles: white space to rest the eyes, visuals.
- Proofread obsessively, many times, using many eyes.

Don't...

- Use jargon (though you may need to mirror back the company's own language to show you're on the same track).
- Use tentative language like we hope, will consider, it seems, may, perhaps, or possibly.
- Obscure the basic message with too much technical detail. Put that in an appendix.
- Drop in boilerplate sections such as company history without tailoring them to the purpose.
- Promise more than you can fulfill: This is much worse than losing a contract.

Even when a proposal request uses complex and convoluted language, your own writing must be clear, simple, and logical.

The letter format lets you shortcut proposals

Formal proposals can be enormously time-consuming for organizations and entrepreneurs. If you're bidding on a government contract or for a big-money project, you usually have no choice but to complete the complex request for proposal (RFP) step by step.

But on many occasions, you can use a simple letter format to save time and anxiety. This approach works well for in-house situations, such as when you're suggesting an interdepartmental collaboration or a new system or program to your supervisors. The letter format is also a good way to go in a number of business situations that don't call for intense levels of detail.

Find opportunities to bypass the whole time- and energy-consuming process of proposal writing by going straight to confirmation letters.

Here's how a letter-style proposal, printed on letterhead, might begin. The subject is creating a marketing newsletter.

Quarterly Newsletter Program:

A Proposal for Denali Interiors

ECN is pleased to present this proposal to custom-create a print newsletter to support all Denali Interiors marketing ventures and meet the following objectives:

- *Build continuity in client relationships*
- *Reinforce the Denali brand*
- *Extend company reach to new prospects*
- *Provide useful handouts for trade shows, conferences, and media contact.*

Notice that audience and goals are built into this first section. The subsequent sections would cover the following:

- Recommended format and visual style
- Content ideas
- Services and fees
- Creative team (brief bios of the principals who'll work on the project)

A close is added to reinforce the project value:

Additionally, as part of your newsletter program, ECN principals will be happy to consult with you on expanding your audience base and ability to reach new clients. We can also advise you on ways to creatively reuse materials we develop, so you are able to make the most of your communication investment.

Even better than producing letter-style proposals, look for situations where you can use letters to confirm agreements—and skip the proposal process altogether. More and more savvy salespeople aim to do the hard sell in person, rather than depending on written communication. This lets them focus their energies on understanding the prospective client's viewpoint and negotiating any barriers revealed in conversation. Then they formalize the deal with a confirmation letter. For example:

Dear Jack:

It was great talking to you about how Acme Provisions can provide food services for your upcoming conference program. We are confident of meeting your specifications on scheduling and quality within your budget.

Specifically, Acme agrees to provide:

(date and nature of service for each event)

The total fee is..., payable on....

All of us at Acme are delighted to have this opportunity to meet your needs.

Your signature below will signify agreement.

You can use the same approach in a variety of situations where you're pitching for business, and practice it at different stages of the selling process. If you meet with a prospective customer, for example, and use your best interviewing techniques to understand what the company needs, you might be able to short-circuit the whole competitive RFP process by writing a good letter explaining how you can meet those needs. Or you can write a confirming letter after negotiating terms in person.

The approach works because selling anything is much better done face to face. People buy from people they trust, so relationship building is crucial. Also, some give and take is only possible in person. You can't do that via printed materials and e-mails.

Even when a bidding situation appears to be totally impersonal, try to find ways to meet the key people and become a real person to them or, at least, build a telephone connection.

TRUTH

33

Root grant applications in "mission"—yours and the funder's

Creating a grant application is similar in many ways to a business proposal. The specifications provide major clues on how to articulate your goals, and the tone and language to use. So study them in depth.

A good grant request feels rooted in an organization's mission. It must also tie closely into the mission of the funding organization, because advancing that mission is the decision makers' own goal. So here, too, know your audience. Research the funder's history and previous grant recipients. Figure out what the funding organization values: Intra-agency collaboration? Groundbreaking ideas? Programs that promote institutional change or help for specific populations?

Successful grant applications focus on goals, not process; what will be accomplished, rather than how.

Foundations and donors understand that maintaining day-to-day operations is a constant challenge for many nonprofits. Nevertheless, they usually prefer to fund programs—projects and initiatives that will be created, in whole or part, with the money they give. Funders these days also like to see nonprofits collaborate with other agencies, public entities, and corporations to extend the scope of their own capabilities and operate more efficiently.

A good grant application tells a good story. The natural story line suggests how to organize the request, though of course you should follow the format you're given. One way or another, try to...

- Establish a need.
- Propose a solution.
- Explain how the idea will be implemented—process, partnerships, collaborations, staffing.
- Demonstrate the organization's solidity, strengths, and track record.
- Present the budget clearly.
- Outline what the program will accomplish and exactly how program success will be measured.

Good grant-writing techniques—Applications are evaluated solely on the basis of what you write—so use your best skills to produce a concise, clear, readable, jargon-free, and error-free document. Editing and proofreading are essential. Spelling mistakes and bad grammar undercut credibility.

Aim for the right level of detail to support your case. If you are using boilerplate pieces—such as a standard description of the agency—take the time to tailor them to the application. If you have a typical committee-written obtuse mission statement, don't use it; or if you must, tuck it somewhere inconspicuously, and where it counts, explain your organization's reason-for-being in simple, direct words. This may not be easy, but it matters.

An executive summary is usually a plus. Take the time to work out a strong, concise overview of the story you're telling. It should set up the reader to view the application in the context you'd like. Ideally, try to excite the reviewer.

In each section, make sure your important points are right up front. Invariably, grant evaluators work through a pile of competing applications, and it is wearying work. If your meaning is dense, buried in excess wordiness, or hard to follow, your application joins the reject pile.

Most applications are very repetitive, and this can make a judge especially grumpy. How much material should you repeat in the various sections? As little as possible. If you're applying for a government grant, or any with word-count limitations, you usually can't afford the space. Even when there are no limits, it's risky to bore your readers. They'll just stop reading. Include just enough repetition to make each section self-contained, in case different readers evaluate them. But unless you're applying for a very high-end, big-money grant, assume a series of readers will each review the entire application and score it that way.

Focus on goals, not process—Many applications and proposals talk about what a program will "do," rather than what it will accomplish. In marketing terms, it's a matter of "features" versus "benefits. For example:

We will present three workshops weekly for 12 weeks to 15 teenagers.

Works better as:

We will teach 15 sophomores to be peer mediators, able to intervene in school conflict situations and in turn to train dozens of other students in the techniques. A better school atmosphere will evolve every year.

Another example:

The funding will be used to buy nine horses so we can expand our riding program.

Doesn't work as well as:

The nine additional horses will enable us to give dozens of physically impaired children the life-changing experience of freedom and power their limitations generally deny them.

Individuals and organizations give money to make the world better or to make life better for people. Your application should show how you will do that.

And, provide as much hard evidence as you can of your organization's success, and, how you plan to measure the effectiveness of the proposed program. Today's keyword is *accountability:* It can be hard to show how you've changed people's lives, or what bad things you've prevented, but those who make decisions on substantial grants want applicants to talk the language of business. Find ways to quantify "return on investment."

Build relationships, too—Giving money involves trust and faith. So relationships make a difference. Look for opportunities to personalize your dealings with grant givers and make your organization more than just a name to them. You might call to ask whether your agency is eligible; pose questions about the best way to fill out the application; find out who previous recipients were, if that information is not easily available; check whether the application was received; or if you have something important to add after you've applied, like another grant award. Communicate the importance of the agency's mission and your total belief in it as a subtext.

Follow up, too. Be gracious and appreciative if you don't get the grant. Consider asking what you might do differently in the future.

And puh-lease, when you get a grant, supply a solid written report on what you accomplished with the money, even if the funder didn't require it. Surprisingly, many nonprofits don't do this, or do it poorly, which shows disrespect. If you apply to the same organization again, this omission will be discussed by the review committee and held against you. Count on it.

TRUTH 34

Writing is the missing factor in your competitors' Web sites

Most organizations create Web sites with an underlying flaw that can be fatal, virtually speaking. The site is painstakingly structured and designed, and when everything is in place, someone says, "Okay, now go get the copy."

Then the PR staff digs out the company's old promotional brochure so it can be crammed into the spaces left by the designer. This presents two problems: First, even good copy written for print is unsuitable for online use. Web site copy generally should be half the length of print copy and devoid of empty phrases and promotional hype. Because reading on screen differs from reading on paper, to work well, Web site copy needs to be carefully crafted for the medium.

And second, by leaving the writing for last, chances are that no one has planned the site properly. The designer cares about how the pages look. The programmers are concerned with site structure and navigation. Other specialists may focus on search-engine optimization. Who asks what the site's goals are, who its audiences will be, what features might draw visitors, and how they will use it?

Even if the site development team includes marketing mavens, it's important to give the writing function a place at the table during the early planning stages.

When you're involved in building a departmental Web site—or in creating one for your own business—fill the vacuum: Assume the role of thinker and planner, and view the job as packaging information. Look at the design and production functions as serving the business purposes you want to achieve. Do this tactfully, of course. This is a team enterprise, and it is the good meshing of skills—and planning—that creates great sites.

Plan your Web site before you structure or produce it, using the writer's big-picture perspective. This alone will make it rise way above the cyber-crowd.

What defines good Web site writing?—As always, knowing what to aim for makes the difference. But for Web sites, "writing" means more than just contributing words. Whether you're able to work with the best graphic designers and programmers or are in do-it-yourself mode, Web sites require a big-picture perspective.

Good Web site writing is built on how people look at sites and use them. Remember that they are viewers, or users—not readers. They scan. People decide what information they want and then dive for it. Watch a high school student work on a paper on, say, the early life of Theodore Roosevelt. Probably he or she will Google "Birthdate TR," and pull this piece of information from the top site or so listed; then it's back to Google for "early education of TR," "TR's parents," and so forth, checking only the first few sites referenced for each case.

Good Web writing is tightly written—It's not edited down but crystallized and stripped of everything the user won't find compelling. Hyperbole about a product, service, or person...a promotional tone...empty statements are turnoffs, and users will skip them or go elsewhere quickly. The short-everything-is-better rule applies doubly to Web sites: short, basic words...short sentences...short paragraphs one or two sentences long, three at most. People don't like to read on screen and will resist scrolling, so you must give them less.

Think about information chunking and graphics—Package the information for the audience: Use self-contained chunks of content within a clear context so that people know where they are and what they're looking at. Individual pages should be self-explanatory—even if that means repeating material from other pages.

Because people usually come to your site via links on e-mails, other sites or search engines, they will see only what's called a "landing page"—the page that contains the material they want to see. That's why most visitors will never see your home page or read through the site sequentially. So aim for clear labeling—descriptive heads, subheads, bold lead-ins, color to draw attention. Lists and bullets work well.

Provide continuity elements throughout the site—Even though we're learning to think in self-contained information blocks, you should still provide for continuity. The graphic look should be consistent. Try to end each content page with a suggestion of something else the viewer should look at next—another product, how the company works with customers, or ordering information, for example. Some specialists feel that as many pages as possible should end with a call to action: Buy now, call us today, subscribe to our newsletter, or ask for our brochure.

Think links—Based on your subject and purpose, think how one page or one chunk connects to another, or to an outside source. Make that connection for your viewer by hyperlinking, providing instant connections to another part of the site or an online resource. But too many links will interfere with the visitor's ability to absorb content. Also, if you're linking to lots of information on other sites, your viewers may never return to your page.

Choose your writing style and tone carefully—Good Web writing can range from fairly formal to pretty casual, depending on the business. A law firm can't be flippant, for example, but an advertising firm may consider this tone a definite option.

One way to make your tone less formal is to use contractions (*it's, won't, can't,* and so on), but if you're aiming your site at a global audience, go easy on these because they may be misunderstood by readers whose first language isn't English.

All sites should feel friendly, accessible, and as conversational as possible.

Should a company site look and sound consistent from page to page—even if individual sections are administered by various departments? That's a company policy issue, but visual variation can work—as long as there's graphic consistency and the differences aren't extreme enough to confuse viewers.

On the other hand, word usage, spelling, punctuation, and terms and acronyms should be consistent throughout the site. Even if various pages of your site were created by different departments of your organization, the pages should be consistent with regard to the basics.

35

Web sites built on keywords and content build traffic

How do people find their way to your site? The same way you find a resource on the Web yourself—by typing a word or phrase into a search engine box and seeing what comes up. Therefore, work up a list of keywords and phrases for your product or service. Build your headlines and titles on them, page by page, and incorporate them into each page's content. Three or four times are enough because too many mentions might be penalized by the search engines. This is part of search-engine optimization (SEO).

Suppose you sell dance shoes. Your keywords might be *ballet slippers, pointe shoes, ballroom dancing shoes,* and so forth. But people might type in Danskin footwear, or shoes for salsa, or dance-rehearsal supplies, etc. So you'll have to brainstorm with colleagues or staff, scout competitors to see with what search words they pepper their copy and headings, and spend some time experimenting with search engines to test out candidates and narrow down your list. In fact, if you Google "searchwords," a number of services will come up that will do this work for you; some offer a free trial run.

Some Web site producers build their whole approach on search words. The words are first defined through an elaborate process involving specialized software and services. Then every page is written to incorporate the relevant words and phrases, three or four times per page. It's challenging to make such copy creative, but the strategy can work amazingly on e-commerce sites.

And, be aware that every page must also have an HTML-coded title geared for search engines. You'll want to work with your programmers or Web developer on crafting these.

Your traffic depends on SEO—search engine optimization. Build copy around keywords and aim to give visitors truly useful content based on your, or your organization's, core expertise.

Make content king; the search engines do. They rank sites in part by how much useful information they contain. In planning or expanding a site, think about what will be helpful or interesting to the people you want to attract. Original content is the best route to credibility, too. You might include sections such as these: FAQs (frequently asked questions), e-newsletters, company research papers or reports (perhaps condensed), press releases, unbiased information or news about your industry, articles on how to do or achieve something, reviews of new products in your field, or statistics that show trends.

Aim to be useful and be inventive. For example, in establishing a new medical partnership, a friend realized that there was no common calendar of events in her specialization, so she added one to her group's Web site. With a little assembly work every month, the partners made their site a "must" destination and clearinghouse, which helped build its reputation, visibility, and referrals. A site that sells beads can demonstrate ways to make jewelry; the dance shoe site could talk about how to fit ballet slippers.

Intranets and extranets—Well-designed company intranets help foster in-house collaboration and knowledge/information sharing among employees. They serve the same unifying function for company suppliers, distributors, shippers, and customers.

When you write material that will be posted on your company's intranet or extranet, the same rules apply as for regular Web writing: Write short, write clearly, compose brief paragraphs, use hyperlinks to move readers around the site, and make the column width narrow—probably 60-characters at most. Use headlines, subheads, and other devices to break up material.

Some more Web site do's and don'ts

Do understand your target audiences really well and orient every element to them. Try for simplicity. People like sites whose elements don't fight each other for attention. **Do build in multiple ways** to find the same information, because different people search differently. On a clothing designer's site, for example, one visitor might look at the menu at a page's bottom to find scarves and another might check the product listing on the home page, or notice a linked reference on the designer's "About me" page.

Do repeat information so each page is complete, remembering that users won't read everything and won't view pages in sequence. **Do match your menu items to your pages.** If your menu bar says "Business Technology Strategy" but the page that opens says "Financial Management," rethink your menu or page labels.

Do build in ways to update your site, whether with news, new product information, articles, and so forth so that you can keep it fresh and keep your audience coming back. **Do use the say-ability test:** How does it sound when read aloud? **Do humanize and build empathy** whenever you can. Show real people using your products or services, for example. This applies particularly to nonprofits.

Do use visuals that serve a purpose and relate to the content, such as photographs, illustrations, or graphs. **Do use graphic tools** to direct the user's eye—typography, color, icons, design. **Do offer ways for viewers to get more information:** a person and phone number to call, material to ask for, links to technical specs, or a working e-mail link to someone who'll answer a question.

Do proofread obsessively. Millions of people may see your mistakes and laugh! Check that every link works, both those that direct viewers from one part of the site to another, and those that connect visitors to off-site resources. Check this periodically: People are annoyed when they're referred to a page or an entire site that no longer exists.

Do end each page with a call to action. "Call for an estimate today," with a phone number, for example—or with a lead-in to another page. For example, "If you like our earrings, check out matching necklaces," and link to that.

And some don'ts: Don't overload the senses. Visual and aural glitz, and meaningless motion, are distracting and slow a site down. **Don't write copy that is cute or clever;** many people have no patience for it. **Don't use questionable humor**—which is almost any humor to somebody. **Don't put a mission statement on the home page**—or anywhere else if you can help it—many epitomize everything Web site writing should not be. **Don't put up a Web site and forget about it.** Nothing is more off-putting than evidence that nothing's been changed for three years.

TRUTH 36

A home page must crystallize who you are

To demonstrate how you can go about creating a Web site structure and home page, let's plan a simple one for a craftsperson who's starting from scratch. And let's suppose she creates a range of crafts and clothing that sell from $50 to $500 and have a "Vermont" theme.

Goals—To sell products via the Internet and bolster the owner's artistic credibility when she sells at fairs.

Audience—People who like handmade crafts and the Vermont "feel."

Content plan—Translating these into the Internet medium, she will want, minimally, the following:

- Home page
- Meet the Artist (a version of About Us)
- Product showcase (one page or set of pages for each kind of product with brief introduction)
- "Where I Am" (updatable page on craft fairs where the artist will show her work)
- Shopping basket to hold selections
- Ordering capability
- Contact information
- Content (such as an article on how handmade fabric is woven)

Even though many prospective buyers will go straight to a particular product page and never see the home page, nonetheless it sets the stage for content, structure, and style. So, how to write a good home page? Here's how our artist's print brochure reads:

Introducing Laura Jones:
A Craftswoman in the Best New England Tradition

Vermont artist Laura Jones is the creator of unique, handmade crafts and artistic pieces that embody themes of her native Vermont landscape. Her highly original work is in the collection of the New England Museum of Modern Folk Art...etc., etc. She has taken blue-ribbon honors in competitions from Maine to Texas.

Her wide range encompasses painted textiles, wall hangings, clothing, and jewelry, all inspired by the Vermont world in which she was born and raised. Her artistic sensibility has been honed by both education and experience....

Even if you don't think the copy is awful, it's nowhere near right for a Web site. Who'd sit still to read it? How to cut to the chase? Here's one way:

Laura Jones Studio
One-of-a-kind crafts and clothing with Vermont themes

Wall hangings
Painted fabrics
Painted scarves
Painted ties
Enameled jewelry

Unique gifts and collectibles made by hand—buy direct from the award-winning artist's Montpelier studio.

Unless you're Microsoft or Google, your home page should include whatever it takes to instantly position your organization.

Each listed product links to an inside page. The items outlined under Content Plan can be arranged as a menu any way the site owner wants or a designer suggests.

Far more complex sites can be planned out in a similar manner. The home page must reflect the best structure for the site's purpose and should be worked out by the full site-development team. Software has evolved that makes content changes easy, but changing the site's structure can be harder. Ideally, a good site plan takes future development into account.

For easy navigation, organize logically. It can be helpful to visualize the site as a tree branching off into roots, which lead to a succession of smaller roots. In the case of the artist's site, for example, you reach a Product Showcase page by clicking on that link. That page in turn can link to wall hangings, enameled jewelry, and the rest. The jewelry page can lead the viewer to earrings, necklaces, and so on.

Alternatively, the visitor can click on the direct links from the home page to each category, or might find the earrings page directly from Google by searching for "handmade enameled earrings."

Explaining the site—The artist's home page features a prominent tagline:

Unique gifts and collectibles made by hand—buy direct from the award-winning artist's Montpelier studio.

Many current sites don't bother to define or explain what the company does or is. As a result, you may find yourself on home pages that make you figure out what relation the organization has to what you want to find, or the basic nature of the site itself. Unless you're Microsoft, it's best to tell the visitor whatever is needed to "position" your site. However, the statement must be very concise and as strong as possible. Don't be surprised if it takes a lot of thought to do this right. The exercise forces you to drill down to your core marketing message.

Cross-promote like crazy—Link your Web site to all your company blogs, and vice versa. Offer print materials about your organization and specific products and services. Link to your e-newsletter, or offer subscriptions to those who fill out a brief form (great for building your database). Connect with podcasts if you can provide those, and video you are showing on YouTube. Promote your upcoming personal appearances as a speaker or consultant, or your company's presence at a conference or convention, and link your site to your social media profiles. And of course, emblazon your Web site address on everything you produce. One of us recently received a small piece of pottery, and the company's URL was etched on the bottom. That's smart.

Worth the trouble?—Even with the services and templates available online to facilitate Web site creation, or the support services available in a large organization, good Web sites are very challenging to produce and update. But it's hard to imagine operating any kind of business without one these days. Surprisingly, *The Wall Street Journal* has reported that only 36 percent of small businesses—defined as having fewer than 100 employees—have Web sites.

If your organization doesn't have the best Web site it can pull off, it's missing major chances to reach audiences that would otherwise be unreachable—all over the world—every hour of every day.

TRUTH 37

Blogging and social media are powerful business tools

Blogging has transformed our world. The blogosphere is where we're likely to go for information, ideas, and advice in every situation, from how to paint the living room to how to find the best hotel in Timbuktu and what to do when we're there. Today, many people trust bloggers more than they do traditional authorities. Information and ideas have become "democratized" and interactive. We can all be part of the conversation with unbelievable ease, using software and services that are free or inexpensive.

For corporations, blogs have become must-have vehicles for selling products and services on a more personal level, and many encourage employees to blog about the company. For professionals such as lawyers, accountants, and consultants, blogs may be the best way ever invented to establish themselves as authorities, reach clients and prospects directly, personalize relationships, and become known in their communities of choice.

Blogging has changed the political landscape, too. Blogs can and do bring down politicians, generate controversy, expose injustice and corruption, find contradictions and discrepancies, and reveal facts that at times seem inappropriate for public consumption.

The exploding use of social media extends the blogging revolution. Some experts predict that e-mail will soon be another has-been communication form and that Web sites will morph into blogs.

This means that when you blog, you're competing against thousands with the same interests and self-interest, so doing it well can make a big difference. The concept of "build it and they will come" can really work. Provide something of real value, and you'll be found.

Good writing strategies work in all digital media but must be applied thoughtfully to each individual communication channel.

Here are some points that apply to social media as well as personal and business blogging.

Remember that when you use these media, even when privacy options are available, you can't really keep your professional and personal worlds separate. They have merged. A prospective client or employer will check you out on Facebook, MySpace, and LinkedIn and also look at your postings. And, like e-mails, blog postings never, ever go away: They are archived and can be redistributed endlessly.

Remember also that using these media effectively can be very time-consuming. If you want to build a good blog with a following, rather than just post opinions on controversial topics, you need a steady stream of good material that makes a genuine contribution. Here's how to do it.

Know what you want to accomplish—Promote yourself? Sell a product or service? Share special knowledge or expertise related to your company, profession, or hobby? Or are you seeking to express your opinions and connect with kindred spirits? All such goals are legitimate, but when you define your own, you can be successful. However, keep in mind that blogs and social media postings don't work when they're blatantly promotional.

Know your audience—Define whom you must reach to accomplish your goals. That will tell you what focus might work, what you can offer that will be of real value to your audience, best content choices, and perhaps the right technical level for your material. It's becoming increasingly easy to pinpoint the groups you're interested in through free online services.

Brainstorm content based on goal and audience—A fairly tight focus can work very well so you or your company is identified with something specific that sets the site apart from most others. Specialized knowledge is what drives the Internet. Tie your subject into your own best knowledge base and passion.

What special expertise do you have? Is there an information niche you could fill? For example, a former Wall Street analyst set up a blog that posts financial filings with the government, and it quickly became an essential resource for reporters. An automotive industry executive blogs on developing a new generation of cars, in deep technical detail, and this site attracts thousands of auto groupies.

Links are content, too—Remember that in the blogging world, providing resources that readers can access with a click is valued for its own sake. If you like reviewing scads of materials and can put them into perspective, that's a service. Building up your links is also the way to get read by more people and move up on the search engine lists.

Listen, learn, share—Social media and the blogosphere offer amazing opportunities to monitor the networks and communities you're interested in, socially, professionally, and commercially. So listen to the online conversations and comment only when you've absorbed the protocols of the specific medium and when you have something of value to give. Do your homework and research. Have the patience to build your credibility, and relationships, gradually over time.

Don't assume that only your friends will read what you write on a blog or social media site—Think of all the students who've hurt their chances of college admission by posting questionable pictures on a site, or by telling unsavory anecdotes about themselves. Not to mention job applicants who are totally surprised when an employment manager looks up their Facebook pages and finds something offensive.

During a major trial recently, a key witness was similarly astounded when her blog posts, in which she described her ability to lie and the money she might get from false testimony, were entered into evidence. But blogging isn't a real-life thing, she protested. *Wrong:* In today's world, blogging is as real as it gets.

The principle holds for microblogs like Twitter, too, which gives people the chance to post very short comments of up to 140 characters. (See Truth 40 on writing for microblogs.) Most people think these mini-postings are as casual as you can get. But just ask a PR executive who tweeted about a city where he gave a major presentation to a client based there. His negative comment about the place was relayed to top executives at both the client's company and his own. Not a good career move.

TRUTH 38

To blog for yourself, be yourself, but carefully

Blogging and social media venues are highly enticing to casual communicators, but whether your goal is to build a personal community or promote your own business or organization, don't be fooled. An instant delivery system doesn't mean that dishing up your spontaneous thoughts will accomplish anything for you.

In addition to taking the trouble to write well and self-edit well, here are some ideas for making these media work for you.

Be yourself—Use your own voice, your own personal viewpoint, and definitely, your own name. Choose subjects you're comfortable with and value the fact that these new media allow for enormous flexibility. If you're someone who has carefully thought about a subject and taken the time to read other people's opinions, and you enjoy commenting from a personal perspective, do so. If you think you're interesting or entertaining, blog and microblog away. On the other hand, if you shy away from talking about yourself, that works too, probably even better: Offer practical advice about what you know, share industry trends, solve a problem. That's what most people are looking for.

Writing tone and style—Conversational-informal is excellent for blogging, but don't get sloppy. Despite the fact the blogging looks like a fre-and-easy medium, readers hate bad spelling, lack of punctuation, and run-on sentences that interfere with instant understanding. Write tightly using short words in short declarative sentences that have rhythm. Keep paragraphs short. Cut all the extra words and thoughts that detract from your core message. The say-ability test—reading the piece aloud to find out where the stumbles are—is really helpful in blogging.

Make your tone match your blog site's goal and audience. An accountant might choose to sound analytic and authoritative rather than using a stream-of-consciousness style, but an artist might not. Ideally, your tone and style should reflect your personality.

How long?—Online formats demand brevity. A print article typically needs to be crystallized to probably half its length to be read. If you're posting a viewpoint on someone else's site, just long enough to be interesting works—in the range of 250 words. If it's a business site and the subject calls for it, 1,500 to 2,500 words can be fine—as long as your information and research justify the length. In either case, make material easy to follow: Numbered lists and bullets work well. Use subheads even on short posts and boldface words to draw attention. Build in white space to avoid a forbidding look of density.

Write good headlines—This is crucial. For blogs, construct headlines based on a complete thought or sentence so what you're writing about is absolutely clear. Readers tend to scan only the first few words, and so do search engines and news feeders. Therefore, put the bottom line on top—meaning at the left, in this case. Start with the most important three or four words, even though this can lead you away from the "action" feel you're usually trying for and may sound passive. For example:

Missile defense systems cost us billions in secret budgets

Blogging for profit is hard to do: some ideas

Small business accounting systems can save you millions

Promote discussion—If you want to build a lively blog with different viewpoints, you need to promote some give and take. You might build a question into the content or ask it directly at the end of your own comment: "That's how I feel. Do you have a different opinion?" Whether you allow open access for posting comments or monitor them before publication is up to you or your company. Either way, develop a clear policy and present it on the site.

How often?—Most successful bloggers say they try to stick to a regular schedule and blog at least weekly to keep the site fresh. Operating your own blog is definitely a big commitment, but if you don't keep new material flowing, the time you do put in will be wasted. If your aim is to become part of a community or to build up your own readership by posting on other people's blogs, it's also best to do it regularly. To get noticed, make sure what you post is relevant and well written.

Which social media should I use?—This landscape changes even as we think about it, so take the time to stay up to date and to evaluate what could be productive for you. There's no question that building up an effective presence on social media can be time-consuming. To best leverage the opportunities LinkedIn offers, for example, you need to invest time in adding contacts, participate in discussion forums, answer questions related to your interests, and recommend contacts

Conversational style is excellent for blogging but don't get sloppy—you need your best writing skills and must protect yourself against saying anything that can haunt you in the future.

Facebook and MySpace have become much-used business and professional tools, as have microblog sites like Twitter. Other sites that aim to connect businesspeople with each other are Ryze, Spoke, XING, and Ecademy. To make good choices, be aware of what your competitors and colleagues are using, and how. Then use the techniques in this book to write strong profiles, comments, and contributions. You'll stand out.

Blogging for yourself—Never forget that blogging is the most public of forums, just like social media. Therefore, *don't say anything that would harm your current or previous employer,* or your chances with a future employer.

Don't criticize people by name, because that person may well get wind of it and there may be consequences.

If you criticize a company or product, give legitimate, well-thought-out reasons. Saying that a product is "useless" isn't helpful, but explaining that its battery fails after one hour instead of the advertised five hours is a legitimate criticism.

Be totally honest about your identity. Use your real name, and in everything you write maintain the integrity demanded of journalists. If you have a connection to your subject, say so. Remember the company president who blogged about his own organization and competitors under a false identity? It landed him on the front page of newspapers and an untold number of blogs.

Good business blogging is edgy

Do you need to talk your employer into developing one or more company blogs or into endorsing one you'd like to do on behalf of your department or firm? Sometimes blogging is where a difference in generational attitudes comes up. In many cases but definitely not all, older decision makers may not be immediately comfortable with the idea of blogging. If this is true where you work, and you feel your organization is missing out, you can marshal your best arguments and research what competitors are doing to help make your case.

Point out, of course, how stunningly inexpensive it is to blog. It's also a way for a company to control its own story and go right to the public, bypassing the media in its gatekeeper function. And it's an unparalleled way to go where your customers, clients, and prospects are. But don't undersell the time commitment involved. Present this realistically, or you'll just end up with a big responsibility added to your workload.

Blogs can go where no marketing department can venture: directly to customers, prospects, supporters, critics, and complainers. The interaction can transform how an organization does business.

Here are some ideas about using blogs as a business tool if you want to move your company into the blogging world, or expand its applications:

1. **Share information instantly**—Blogs are superb outlets for relevant news about your company, such as a new product or service, new capabilities, and interesting ways to use a product. You can build interest in a product coming down the pipeline, and reward fans for their loyalty with early news or "inside" information. In crisis situations, blogs can deliver news at "real-time" speed.
2. **Ask your customers or clients for direct feedback**—Focus on a specific area: for example, a company-owned hotel's

accommodations. Additional blogs can concentrate on various other aspects of the business, such as the dining experience, bridal accommodations, and conference facilities.

3. **Respond to input on a daily basis**—At least one person in a sizable organization should be responsible for monitoring the bloggers' universe on a daily basis. Most organizations of any size or significance face blogs whose sole focus is to badmouth them or play "gotcha." As politicians have learned, false statements must be countered immediately. Even negative postings that are true or somewhat true should be dealt with quickly because thousands or millions of people are probably reading them. Your company needs its own instant information outlet—its own blog—rather than just responding to negative attacks on other people's blogs.
4. **Encourage company employees to blog**—Do this even if they might post comments critical of the company. Several major information technology companies promote blogging, either on the company's official blog site, or on the employees' own sites—without vetting the blog posts beforehand. Contrary to the managers' original fears, no company secrets seem to have been revealed. Interestingly, they found that even rantings by unhappy employees offered a benefit, because it gave workers a public outlet for their frustrations and sometimes led to workplace improvements.
5. **Directly ask your company's customers to make suggestions for improving your products**—For example, a leading paint manufacturer operates a company blog site that encourages customers and potential customers to ask questions and to tell the company what they like about the various products, and what they don't. Managers have made some useful product changes based on this public feedback. A baby products company created a "mother's panel" to offer advice on new products and needs. That's transformed its entire marketing strategy.
6. **Respond to complaints**—When you receive client complaints, narrow your scope to the issue at hand and always be helpful and conciliatory. Telling customers that whatever happened to them wasn't really your business's fault isn't helpful and will antagonize them needlessly. Blog posts are information sources for other

prospective customers, so answer complaints specifically, and if possible point out ways the situation might have been avoided.

7. **Bridge distances**—If your business has locations that are geographically scattered, employs a far-flung global workforce, or relies on collaboration, corporate blogs are great ways to bridge time zones and distances. Blogging casts a wide net, and posting a question or posing a dilemma can bring you innovative solutions. The informal feel that blogs promote can boost relationship building, too.
8. **Deliver content value**—If you want people to read the blog and to keep coming back, fill it with useful information for your chosen audiences. Solid material draws people and search engines. In-depth articles on specialized subjects are excellent (see Truths 47 and 48 for advice on how to write them).

Tie blogs to your marketing

Track your traffic and rankings and respond to what's working. Promote your blog by identifying and incorporating keywords, just as for Web sites, and submit them to search engines; also submit your blog to one or more blog directories. Link to as many other relevant blogs as you can. To help justify the time you're investing, think about other ways to use the material you've developed for your blog.

Whether you're blogging for yourself or your business, be sure to link your blog to your Web sites, and other blogs, as appropriate. Your letterhead, print materials, and even e-mails should include your Web and blog URLs. Consider your varied marketing efforts as building blocks that create an image, or persona, that's more than the sum of its parts. If you have an e-newsletter, promote it on the blog, and the other way around. If you've issued a white paper or product catalog, promote them electronically as well. That way, you feed traffic to your own media and build an audience.

However, you also need to aim for some consistency in graphic appearance and how you present your organization. That's branding.

TRUTH 40

Tweeting and texting: the ultimate self-edit challenge

It's clear that the ultra-short message, in the form of text messaging or mini-blogging like Twitter, has become a major means of interaction globally and an important business tool.

Texting (short messages sent by mobile devices that use SMS, short message service) is the medium of choice for many teens and twenty-somethings, who use it along with social media for general contact and prefer it to e-mail.

Many older people also depend on it for making plans, emergencies, and on-the-go contact, and it's gaining traction as an advertising medium as well. The real estate industry, for example, uses text messaging to contact potential buyers and deliver new listings instantly.

Twitter and similar mini-blog systems are shaking up traditional media and even the established digital channels. This is despite (or because of) the fact that it limits every message to 140 characters flat out. While many early tweets may have relayed what the writer was having for lunch, twittering is now serious business.

It's a super-efficient way to reach friends, colleagues, acquaintances, customers, prospects, and communities of like interest. And vice-versa: Large organizations are finding that an active Twitter "presence" is essential to branding and customer interface. It can be used to steer an audience to a new product or event and is an instant way to answer inquiries.

In fact, now that a problem can be brought to a company's attention immediately—for example, by a disgruntled customer at the airport whose flight was canceled—it must be resolved just as quickly, because the tweeter can broadcast his or her unhappiness to untold thousands of other people.

So if you work for a substantial organization, Twitter probably belongs in the communication toolbox. If you're an entrepreneur, a professional who wants to establish expertise, a consultant, or anyone

If you tweet for business, an overall strategy is far more productive than random messaging.

who wants to engage with people or groups, Twitter can be great for your purposes.

So how can you effectively write for short-message media?

Don't be disappointed to hear that virtually all the tactics you're learning in this book apply. Super-speeding the delivery system does not mean you should short-shrift the writing process when you're messaging for business purposes. On the contrary: Look at texting and tweeting as the ultimate self-edit—a sort of final writing exam.

Think the 140-character limit is tough? Consider that at least one sophisticated cook is disseminating recipes via Twitter, others are reviewing movies, and some well-known writers have started to distribute stories in short Twitter bursts.

Successful tweeting needs a plan—If you tweet for business, an overall strategy is far more productive than just random messaging. Know what you want to accomplish: Promote a service? Establish your credibility? Strengthen friendships? Exchange ideas? Be part of a community?

And know the audiences you want to reach—what interests them, what works for them. As with every other writing medium, substance counts. People value information that will benefit them much more than casual entertainment, though well-delivered material always works best. So keep to what you really know.

It's useful to view Twitter as a supplement to your online presence. You can use the 140 characters to point people at your blog or Web site, which can offer more substantial information and ideas.

Make your writing as brief, clear, and conversational as you can—Use short words and simple sentences. Twittering is the antithesis of the old one-way communication approach, so try to promote dialogue. Questions are good, for example: "What do managers hate most about employee writing?" This is a good way to draw readers to an article or blog post, too, whether yours or someone else's.

Lists are effective: "4 ways to say 'thank you' for a client referral." "3 ways to roast turkey for non-cooks." These examples would all continue with the linked URL, preferably condensed through tinyurl.com, bit.ly, or another service.

Strong headlines that feel urgent or essential get your tweet read and re-tweeted. Example: "6 Rule-Breaking Ways to Pitch Creative Services." If you want readers to forward your messages, edit for clarity and directness. It's okay to use contractions, but not the abbreviations common to texting. Do use good punctuation and capitals—they cost nothing.

By the way, if you aim to be re-tweeted, stick to a maximum 120 of those characters, leaving room for the re-tweeter's message and identification.

To launch yourself in the Twitter world, as with blogging, first listen carefully to those already in a group that interests you. Ever notice how socially adept children join a game or activity? First they observe silently from the sidelines, and after a while make comments about the action. Then joining in and becoming a player seems natural.

But: Never send tweets to foster controversy that could backfire, or criticize other people in abusive ways. You might end up publicly apologizing to thousands of people. How humiliating is that?

About texting—Should businesspeople use abbreviations when they text? Fast answer: Allow for differences in levels of texting savvy. If you're not sure your audience will understand your message, spell things out more thoroughly. If there's the slightest chance of offending or confusing the recipient, try to condense your wording without using acronyms and other shortcuts. And you can text that you'll telephone or e-mail the person at a specific time.

Another question: Should constant texting by "hyper-socializing" young people be accommodated in the workplace? For that matter, should constant checking for Facebook news be condoned?

The jury's very much out on this, but the real question may become, is there a realistic way to stop such activities should a company or manager want to? What's certain is that communication channels are morphing—and in turn, are changing our business culture in profound ways.

41

E-letters focus marketing and reinforce branding

Print newsletters are generally complex, time-consuming, and expensive to produce, but the e-revolution makes it relatively easy to create and deliver e-newsletters.

Effective e-newsletters can run anywhere from a few paragraphs to pages of copy, and well-done ones can be very productive for a business, a department, or an individual. But they must be well planned, well written, and useful in some way. Sure, it's easy to push a button and distribute your communication to hundreds or thousands of people, or have a distribution service do it for you. But there's little point in bringing yourself or your company wide attention with a document that undercuts your professional image rather than enhancing it. If recipients feel you're wasting their time, they simply press Delete and you lose.

E-newsletters build relationships and keep your client base connected to you. The technology is easy, but to do a good one, you need a goal, an audience, and something to say to it.

This means that you should not undertake an e-newsletter unless you're prepared to put time into writing it or have help. Your newsletters must have substance and be simple, direct, concise, very readable, and free of jargon. E-newsletters are worth doing for the same reason that print newsletters remain important: They build relationships. Sending periodic issues keeps your client base connected to you so that when they're in the market for your product or service again, your firm will be the first that comes to mind. For prospective customers, a credible flow of information is a superb selling tool. Further, newsletters give focus to your marketing efforts and reinforce your branding.

There are innumerable ways to approach e-newsletters. Here are some general guidelines to get you on the right track.

Know your audience and have something to say that will interest or benefit its members—The content may be timely news of your industry, advice, information on how to accomplish something they may want, or ideas. You might aim to provide a resource that connects recipients with multiple sources of information via links, or with your own in-depth resources. Brainstorm with colleagues to come up with creative ways to draw on your expertise in a newsletter your audience will value.

Know your goal—Do you want to support the marketing of a product or service? Make people feel good about your organization? Raise money? Promote recognition of your name or brand? Draw people to events or involve them?

Organize clearly—People don't read online material; they skim or scan it. To make your newsletter scannable, use bold type in your text, and prominent subheads, to call attention to major elements of your message so readers can instantly see what's of interest to them (or not). If your newsletter is longer than one online page, have a list of the articles up front linking to the content that follows.

Don't come across as promotional in all or most of the content—People will not spend time reading anything that looks like self-advertising in this format, unless they are already loyal customers or fans. Give them information they will value for its own sake, and promote your product or services in a separate section. Special offers or announcements of new products are a legitimate part of the mix. A section can be headed "The latest products from XYZ," or something similar.

Make it look good, but not glitzy—You should have a simple, clean masthead that is recognizable and connects to your organization; well-chosen, readable fonts; and good use of space. It is smart to enlist a graphic designer who works with online materials to create a design for your newsletter. Don't shortchange the editing—nothing works against you more than misused words and marginal grammar. Enlist a second or third reviewer.

Issue the newsletter regularly, but not so often that you irritate people and get earmarked as spam—There's no set formula to follow on frequency; you'll have to sound that out. Once a year doesn't serve the purpose; quarterly might; monthly is better. Although they're simpler to produce than print newsletters, don't underestimate the time that the thinking and writing demand.

Make your e-newsletter save-worthy, providing material that is relatively timeless—Readers will file an e-letter to read later or store it for future reference if the information is solid. Some organizations number the issues—for example, "Jack Smith's Investment Ideas #162"—to encourage readers to collect all the issues. Archiving the issues on your own Web site so they're available to new readers is a great way to build up your original content, too, which is always a plus for search engines.

Focus—For many e-newsletters, covering one idea is usually enough, although some e-newsletters carry a range of content in each issue: news, articles, columns, reviews, opinion pieces, surveys, lists of resources, and services. Substantial newsletters are most effective when centered on a theme that relates directly to the business being intrinsically promoted. An e-newsletter issued by a garden center, for example, covers garden tips, "plant of the week," ask the expert Q&A, a guide on dealing with pests, and upcoming events and special sales.

Tie your e-newsletter to your other marketing efforts—Link it to your Web site, your blog, and other resources you can provide online. Give people a way to ask for more information about the subject or your firm. Provide clear contact information.

Make it easy for people to unsubscribe—Why annoy them if they don't want to be part of your audience? Refining your database is a critical and constant effort, and "bigger" is not necessarily better.

Find ways to make it interactive—Invite comments, share information or ideas, ask questions and get answers, find out what your visitors and customers want, and tell them how to pursue additional avenues for those more deeply interested in the subject. Ask them for ideas about where else to market your services or who else might be interested.

Make it shareable—Provide a mechanism for readers to forward an issue to other people so they can, without effort on your part, expand your database—and, of course, use that database for the rest of your marketing, with discretion.

TRUTH 42

Good PowerPoint is more than pretty faces: It starts with writing

The speaker strides to the podium...turns down the lights...presses Advance...and slide by slide, his speech splashes across the screen so everyone can stare at the written words and hear an oral version at the same time. Later you get a handout that's identical to the slides. A multisensory experience? No, just a boring one.

If this scenario sounds familiar, be warned that it will become even more so. Many business schools are training students to communicate predominantly in PowerPoint. In professional communicators' circles, however, there's a growing conviction that PowerPoint is often poorly and inappropriately used.

Yes, PowerPoint can add a visual dimension to what you present. But there are major caveats: It shouldn't dominate what you want to say. It shouldn't circumscribe your presentation via its built-in limitations. It shouldn't draw attention away from you, the speaker, more than momentarily.

To avoid producing the typical dull PowerPoint, keep your determination to write and think well front and center. Otherwise you shortcut yourself, and the lack of depth shows.

Begin PowerPoint as a writing project. Why? Because we think in words, even when the delivery is visual. You need to know what you want to say before you package it. If you start by creating your PowerPoint "deck," you'll focus on formatting—instead of content. You'll inevitably find yourself adapting the material to the template, adding unnecessary copy or visuals to "short" slides, and eliminating important points from over-packed ones.

Also, when you eliminate the structure that writing demands, you won't have thought through your subject farther than what's displayed on the screen. You'll miss chances to deepen your grasp of the subject and limit your ability to field questions. When everything you know is on the screen, your lack of depth shows. You're stuck with reading your slides.

So, consider developing your presentation through good old Microsoft Word, or scribbled notes, or whatever is comfortable. Then translate this thinking into PowerPoint, and when the format and content don't jibe well, juggle them till they do. This push-pull process generates good presentations. Have you seen a memorable one lately? Ask the presenter how he or she did it.

Here are ideas for making your presentation effective:

Use the step-by-step strategy to plan your presentation, focusing first on goals and audience—Whom will you address, and what do they want to know? What do they already know? What's the central message you want to deliver? What's in it for them? Try to get across three or four main points and marshal your content in support.

Next, map and organize your content, the substance of your "middle section," and pinpoint any missing information. Aim to build a good lead, like an anecdote—especially appropriate for an oral presentation—and a strong close, perhaps summarizing your conclusions.

Write down your plan, whether in Word or on paper or index cards, and then build your deck.

Think of PowerPoint as a support for your in-person presence, not a substitute—For most of your presentation, the audience's eyes should focus on you, not the screen. And you, the speaker, should be looking at the audience, not the screen. People come to see and hear a speaker, live and direct. They need frequent eye contact.

Never project everything you're going to say and then say it—Bored people resist the message.

Sequence the deck logically and build transitions from each slide to the next—Start with a title slide, followed by your agenda—what you'll talk about.

Anchor each slide with a large-type, one- or two-line heading—Use the screen to stay structured, topic by topic, while you provide fuller information, and to present relevant visuals.

Be creative in structuring the story you tell and how you use the deck—For example, you might define the subject and then head your slides as follows:

> *What's the best process for improving...*
> *The best research shows...*

Here's what we learned ...
Here's what we learned from looking at our competitors...
Our conclusions...
Next steps...
Questions?

You might actually have only these statements on the screen, and talk to each point. The plus factor is that by shepherding people through your own process, you build buy-in.

Use visuals thoughtfully—A lot of meaningless motion and clip art can be distracting. When you must put a lot of information on a slide—charts, graphs, and other images—keep it visually simple and logical for the eye. Take advantage of the medium's potential to explain things dynamically. For example, the lines of a graph can move to show change over time, or various elements can be introduced one at a time to make comparisons easier to grasp.

Rehearse!—As many times as you can bear it, practice your presentation aloud, along with the deck. Edit the slides to support your oral delivery and check that they flow right visually. Aim to present without looking at your notes. When you can, use humor, personal anecdotes, and stories to enliven the audience's experience.

PowerPoint as sole record—For better or worse, the medium is gaining ground as sole record of communication within various industries. Many consulting firms, for example, are using it for all in-house communications, even to communicate with clients, and have simply eliminated use of MS Word or other means of documentation for virtually everything. If you're working in this kind of environment, how can you avoid PowerPoint pitfalls?

Don't be a slave to the template. No medium should preempt your thought processes and logic. Work through your message via Word or another traditional writing system. Then translate as necessary to PowerPoint. There'll be more substance behind the words, concise as they must become. Your ideas will connect better, and your arguments will be more convincing.

Hold on to your goal of thinking and writing well in PowerPoint, as in every medium. You need to know more about your subject than you can show on a set of slides. If you shortcut the writing, you shortchange yourself.

TRUTH 43

Strong résumés focus on accomplishments, not responsibilities

Whether you want to advance your professional career, apply for a new job, or build your own business, getting your assets down in writing is critical. There are various kinds of résumés—a consultant may need a "functional" skills-based résumé, for example, and a scientist a detailed curriculum vitae, or CV—but each should be framed to target your goal and audience.

The principle also applies to the newer social media résumé, which can incorporate video, photos, podcasts, and blogs. We'll focus on traditional résumés, but you can adapt the ideas to other formats if they're better suited to your industry and highlight your strengths more appropriately.

A résumé's goal is to get the interview so that you can sell yourself to the interviewer. This suggests that unless you're a physicist or professor, keep it to one or two pages maximum.

Recruiters and staffing people are on the receiving end of a lot of bad thinking. These tips are based on their suffering:

Don't fool with the standard format—Stick to reverse chronological order and keep it simple. (And yes, employers still want to know what you were doing if time gaps are evident.)

Use Microsoft Word or another standard word processing program—Consider creating a Portable Document Format (PDF) version of your résumé so that it retains your formatting during e-mailing.

Don't let typos sneak in.

Don't clutter the layout—Your résumé should be clean and easy for the 30-second capture, so don't vary fonts and sizes.

Don't use abbreviations that are not searchable or common outside your industry, such as "a.m." for accounts management.

Use keywords—Many recruiting firms and large employers digitize incoming résumés, so make them searchable and build in keywords that make your skills clear. To find good keywords, scour the recruiting ad, scan trade magazines, and check the Web sites of relevant companies and professional associations.

What should be at the top of your résumé? Recruiters differ about whether stating an objective works or not, but it should never be fluff.

Cut out phrases such as, "I'm looking for a job that..." or "My objective in seeking this position is..."

Beginning with a brief profile—an elevator speech in print—that says who you are and what your greatest strengths are is a powerful way to open. You'll note that the examples here lean heavily on industry-specific jargon. Yes, we've advised you to stay away from jargon, but résumés are an exception to this rule. Recruiters and employers want to determine quickly how well your credentials line up with their needs, meaning you probably need to use your industry's buzzwords.

Here's an example of an opening profile:

Energetic sales and marketing executive with 16+ years of successful experience in strategic planning, implementation, and leadership of multimillion-dollar marketing initiatives with profit and loss responsibility. Proven ability to analyze markets, target areas of highest return, and develop strategies to attain organizational goals.

And a different example:

Young, profit-oriented advertising professional with 8+ years of solid experience and a successful track record. Well-organized, self-starting team player with excellent communication skills and ability to stimulate increased productivity.

When you create the right opening profile, it tells you what content to focus on in the rest of the résumé. It also tells the reviewer how to read the rest of the résumé, setting him or her up to take the slant you want.

Stress your accomplishments— Employers aren't interested in a laundry list of your former jobs and tasks—they want to know what you've accomplished. So translate your job responsibilities into achievements and quantify your accomplishments. This is worth a great deal of thought.

Employers don't want laundry lists of what you've done; they want to know what you've accomplished. So translate responsibilities into achievements and quantify everything you can.

Rather than saying, "I redesigned a warehouse," say, "Transformed a disorganized, inefficient warehouse into an efficient operation by totally redesigning the layout, saving an estimated annual $50,000 in recovered stock."

Use strong, meaningful words such as managed, conducted, coordinated, directed, and supervised. "I handled marketing issues" is less impressive than "I coordinated marketing campaigns."

"I entered information into a database" sends a completely different image of you than "I coordinated customer databases and upgrades."

Here's a sample list of accomplishments from a marketing executive:

Exceeded *forecast net income by 14% in first 12 months*

Grew *an industrial account base from zero into a segment producing more than $16 million in sales revenue in the first 12 months*

Developed *a formal sales program identifying objectives and strategies, deployment of resources, qualification guidelines, sales incentive programs, and support plans*

Using bullets—They work beautifully for documents like résumés, but don't overdo. Who wants to read two pages of bulleted information? Use them to make separate points, or list examples, but create the generalizations that tie them together and make them meaningful.

Should you include—"*References available upon request?*" No, because it's taken for granted. Should you list your *hobbies or sports achievements* on your résumé? Maybe. If you had been captain of the college hockey team, that might give you a conversation starter. But if you had headed a partisan political club, it might not be a good idea. Should you cite *honors and awards? Memberships?* Yes, if they relate to the position you're seeking.

How about charitable ventures, community activities, and the like? Sure, but call them *"Community Service"* rather than "Volunteer Activities." If you've written articles or books, given speeches or presentations, or taught courses in the field you're applying for, by all means indicate them on your résumé.

Think future—The good part: When you get the résumé right, the phone rings—and even better, you're prepared to answer it. Hammering out your personal statement helps you know who you are and how to present yourself with clout.

TRUTH 44

Fliers are easy all-purpose promotional tools

You may write fliers to promote your business, a new product, a special sale, an event, or even yourself. You may want to get word out that you're available as a speaker, for example; or you may need a one-pager to hand out if you're giving a presentation or exhibiting at a convention. Fliers are an easy and inexpensive promotional tool.

Fliers can be produced on paper and distributed by mail, at conferences and events, or even door to door. They can also be distributed electronically, as an e-blast. An e-blast can be sent out from a Web host or through a company that specializes in doing this. If you work on your own, you can probably send out e-blasts through your own e-mail Internet service provider, but need to know the limits of how many you can send simultaneously. (Some providers offer service upgrades that allow larger e-mailings for additional fees.)

Fliers are ubiquitous, but they aren't always as effective as they should be. Here are some ways to produce good ones, using event promotion as an example:

1. **Put your business or professional association name and logo near the top so readers know where the message is coming from**—If you send out monthly meeting announcements or periodic event invitations, this will set up the recipients to "hear" your message, especially if they already have a relationship with your company or group.
2. **Use a limited number of words and large type**—Fliers are not letters and should not look like them. They're meant to be read and absorbed quickly, and the more words on the page, the fewer of them will probably be read. Appearance is important, so if it's a do-it-yourself project, spend time experimenting on your computer or get help from a graphic-minded friend. Aim for lots of white space, short paragraphs and sections, and not too much variation in type. Use headlines, subheads, and boldface to enhance readability. And, of course, apply your best tight-writing skills.
3. **Put the bottom line on top so that it won't be missed**—For example, if you're producing a flier inviting people to attend a company trade show, make the name of the show and the date

the most prominent words on the page. The date should include the day of the week.

4. **Create a strong title or headline**—It should be specific and as compelling as possible. "Announcing a Trade Show for the Magic Business," for example, is less effective than saying:

   We invite you to

   THE FIRST TRADE SHOW FOR MAGIC-MAKERS AND RETAILERS

   THURSDAY, APRIL 9 AT NOON

   Another example: Rather than titling a flier "Price Reduction at A-1 Hardware This Month," it's more interesting and specific to say:

   OCTOBER SPECIAL:
   IT'S FALL CLEAN-UP MONTH AT A-1 HARDWARE

   Take 25% off all leaf blowers, gardening tools, rakes, and more

5. **If you're promoting an event, describe it briefly but in terms your reader can identify with**—Remember the "what's in it for me?" principle: Why should the audience care? Tell people what they will learn or how they will benefit from what you're promoting. Questions can work well. For example, if your event is a trade show, "Want to learn the best tips for selling more widgets?" is more effective than "We'll give you tips for selling your widgets." If you're selling yourself, think in terms of, "What problems can I solve for you?" If it's a new-product flier, "What can this gizmo do for you?" should be addressed.

6. **Include all the important details**—These include how and where to register, the cost of the event, and either directions to the venue or where to find directions to the venue. These items can be shown in smaller type, but they're essential.

7. **Include a telephone number that people can call to get more information**—Add a Web site address if the site will list the event or if you want people to know more about your organization.

Address "what's in it for me" by speaking directly to your reader at all times: "We invite you to...," "Here's what you'll learn...," You'll save money...."

8. **If you're sending out an e-blast, be sure the subject line is clear, self-contained, and has a must-read flavor—**For monthly meeting announcements, for example, you might include the name of your organization and the topic: "DMA December: John Ash on the 10 Best Ways to Find Customers." The object is to get people to open the e-mail, of course, and there are various ways of doing this. A follow-up e-mail could have the subject line, "Last chance to register for John Ash on Dec. 12."

And perhaps most important of all: Speak directly to your reader at all times. "You are invited," "We invite you to join us," "You'll find out...," and all the variations, according to your purpose, will focus your flier properly.

TRUTH

45

It pays to think PR and send news releases

If you run your own business or consult for a living, add news releases to your repertoire. They can earn you editorial coverage that is a lot more credible than advertising, and moreover, you need only invest a little time.

Writing releases is a skill valued by most professional associations, too. Practice your know-how on behalf of a professional group, and you'll fast become a valued member. What's worth a release? First and foremost, have something newsworthy to communicate—a story. Many events are worth capitalizing on.

Look for opportunities to announce a new product or service, significant staff addition, office relocation, business success, and awards or honors for yourself, a staff member, or your business. You can also announce an appointment as a nonprofit board member. Through press releases, invite the public to attend a special meeting or event, and invite readers to join a contest or competition. Use historical milestones, such as an anniversary, to gain attention for the company, and give the organization a human face by, for example, publicizing good deeds of the company or employees—food drives, donations, toy collections, work in food pantries, scholarships, and so on.

Moreover, you can generate information to give you a reason to write a release. For example, you can collect some statistics or conduct a simple survey—which can be done via e-mail—on something relevant to your industry. If you sell bicycles, you might survey your customers about whether they wear safety helmets. "Local survey finds that only 38 percent of bike riders wear helmets" makes a good story that might be picked up by a local newspaper. This gets your business's name out there.

> You need something newsworthy to communicate, but it doesn't have to be earth-shattering. Events, contests, anniversaries, awards, donations, a new account, a milestone—all are worth announcing.

Scan your business regularly for connections to the news or things that might interest media and readers. You can identify seasonal tie-ins or find opportunities that offer human interest or visual potential. For example, a children's toy store runs a fashion show for dolls; the young owners dress them up and do "runway" commentary. The owner's simple release drew the editor of the community newspaper to cover the event and take pictures.

Look beyond the obvious to figure out what will interest people. What's special or unique about the event? With a product or service, the most promising angle lies in what it achieves. A new piece of medical equipment may not sound important in itself, for example, but if it helps people monitor their blood pressure more easily, you've got a story to tell.

Most publications prefer to receive releases by e-mail. To get it read, use specific, informative subject lines. With social media releases, make information readily usable for online bloggers, reporters, and editors by chunking (breaking content into short, easily read pieces); and more findable by search engines via "tags," words that identify the content generically. Whatever the media target, always try to address the release to the current editor.

Plan the release

Start with the *goal.* Generally, it's to gain favorable publicity for your business and raise its profile, to support marketing. Consider *audience:* While you're aiming to reach the general public or the segment that might buy your product, you must first win the approval of the "gatekeeper"—the editor. Since the editor's job is to find interesting stories for his or her readers, make your release relevant to the publication's audience.

Preferred writing style—Try for objective-sounding, third-person, newspaper-style reportage. Begin by crafting a good headline, one that crystallizes the message and captures attention. Use short, crisp leads that appeal to the readers' interests. Write in "inverted pyramid" fashion: Start with the most important facts and work down to what is least important.

Use simple, straightforward, "sayable" language: short words, sentences, and paragraphs. Eliminate most adjectives, all jargon, clichés, and empty phrases. Use active, lively verbs. Minimize the number of words ending in *-ion* and *-ing* and the number of times you use the words *on*, *of*, and *to*. Edit with particular attention. Publication gatekeepers notice every spelling error, and they care.

Content and format essentials—*At the top:* Say whom the release is from and include complete, reliable contact information, including your cell phone number.

Body of the release: Cover the journalist's core questions. For example: *What* is the event? *Who* is the sponsor? *Where* and *when* will it take place? *Why* is it newsworthy? *What's* the event's purpose? *How* can the editor get more information?

Keep the release short: In print format, one page total is preferable, and one and a half pages are okay. Your goal is to provoke interest. Keep paragraphs short and use space between them to make the release more readable.

At the bottom: End with a statement that briefly says what your company or organization does; include your Web site address. If you're trying to get a reporter to cover the event, or are inviting the public, include brief, clear directions.

If you have a good photograph: Including it multiplies the chances that a release will be used. To lure coverage, specify the photo opportunities and when they'll occur.

If you're trying for television coverage: State the visual element and exact time frame.

Use the results—Circulate a hit wherever you can: Enlarge and frame it to hang in your office, post it on your Web site, print it in your newsletter, mail or e-mail it to your customers and prospects, send it to your professional associations (who may rerun it in their newsletters) and to any group that might help your cause or product. Consider running an ad incorporating the coverage.

Some of the sharpest PR people we know see events as excuses to create great press releases. They'll even create or adapt an event to justify a release and media interest. It's a good, entrepreneurial way to think.

TRUTH 46

Writing articles boosts your career

Why write articles? Ask Ben. He's a lawyer who specializes in tax strategies for small businesses. An editor overheard him talking about his favorite subject at a party, and asked him to write a brief article on tax questions for her regional business publication. The article evolved into a monthly column on taxes for small businesses. The pay? Zilch. But Ben is reaping big benefits from establishing himself as his region's leading expert in his field.

Ben was lucky, but the fact is, there's a world out there hungry for content. Print and online media just can't keep up with the demand to fill their actual or virtual pages. Consider trade and professional magazines; association newsletters; newspapers (dailies and weeklies); local magazines; pennysaver-type publications; national, regional, and local publications for parents, art lovers, antique collectors, tourists, boaters, and so much more.

Most of these publications also publish online, demanding scads more material, not to mention all the resources that are virtual-only. It adds up to an insatiable appetite for new material. So, when you don't care about the money or a byline in a competitive newsstand magazine, your chance of getting published and making yourself an authority in your field is surprisingly good.

Where do you start?—With what you know. Maybe it's one main thing—like auto insurance, or beauty products, or new technology. If you've worked up any solid, specialized knowledge, you might write for fellow professionals. Or you can adapt your expertise to "consumer" interests. For example, a dentist could write for a local publication about the best oral products for home use. An interior designer could advise readers on good color schemes. A landscaper can write about how to make better use of shrubs around a house or how to maintain gardens without pesticides. A banker might explain how to teach children to budget.

Together, traditional print and newer online media create an insatiable appetite for authoritative content. If you plumb your expertise, you can find markets that reach the audiences you want.

You might also have hobbies you can write about, in which case you can and should include your company or professional affiliation with your byline or bio. Next, research for opportunities that line up with your list. Be open-minded. An editor we know left her job at a big-circulation daily newspaper but offered to write a column about pets, her favorite subject. To everyone's amazement, it was a hit with readers and never seems to run out of subjects.

It's better to "pitch" one or more specific article ideas to an editor rather than just saying, "I'm an authority on such and such; may I write for you?" Brainstorm about ideas you'd like to see a publication cover, a national trend you can bring down to the local level, questions you're most often asked, an event you can illuminate, and subjects you yourself would like to know more about.

Another way: Start with the publication you want to write for. Study three or more issues to understand its viewpoint, its editorial niche, what it has covered lately, and the number of words a column usually runs. Once you're tuned into the editorial thinking, ideas will come. Professional freelancers use this method.

Writing your article for print media—Let's figure out how Ben could have approached his article on the new tax law for small businesses. He knows the audience will have varying levels of knowledge and that he needs to explain why reading this article matters, what the law says and means, implications for small businesses, advice on how to work with the law, an example, and how to get more info—ask me!

The lead—Professional writers spend a lot of time crafting good leads because rarely do you have a captive audience. A lead must capture interest, represent what is covered, and make it appear worth reading. Here are a few techniques journalists use that work well for business subjects. Adapt them as models—some smart professionals keep files of favorite leads they run across to adapt as needed.

The rhetorical question:

Will Regulation 444 help or hurt small business?

The new law will govern how small businesses can deduct expenses, and this April, everyone is expected to meet the new guidelines or face stiff penalties. Here's what you should know.

The anecdotal lead:

Ann March, owner of a women's boutique and my client, couldn't believe it when she'd finished a run-through of her taxes for the last quarter. Returns and expenses had been similar to the last year's third quarter, but she owed $4,200. What was wrong? she asked. In fact, she just didn't know that...

String of declarative statements:

Since 1994, small business owners have used a simple calculation to calculate the year's expenses. Record-keeping was also relatively simple, and the required backup could be loosely interpreted. On January 1, all that will change. Regulation 444 will come to life. Every small business owner will be directly affected.

Direct quote lead:

"I'd calculated my taxes in advance the usual way, before I knew about Regulation 444," says Mary Green, owner of a three-employee local business, MG Clothing Repair. "Then I got the new forms and figured it out again," she says. "I absolutely couldn't believe the difference."

Surprise/interesting fact lead:

The Internal Revenue Service looked at its own books two years ago and concluded that the government was losing $400 million per year because of the way small businesses are allowed to calculate their deductions. The IRS invested in planning a new system—and in January, small business owners all over the country will discover the result.

The rest of the article—Where do you go from here? Just as with e-mails, write the middle, with all the basic information, following through on your content list, logically sequenced; and as you do this, keep an ear out for a good ending that circles back to the lead. For example, in the case of our invented subject: "Understanding the new regulation and using these strategies will minimize the new law's impact on your business—and may save you quite a bit of money."

TRUTH 47

The virtual world offers self-publishing power

Today, we all have the power to be publishers, which presents an extraordinary set of opportunities. If you're a businessperson, consultant, or entrepreneur, posting articles on your own Web site or blog is one of the most effective ways ever imagined to build a reputation, position yourself as an authority, and support the marketing of a service or product.

To search engines, an article with useful information is "original content" that moves your site up the rankings in its category. This means it will be found early on the list that comes up when a search term is entered. (Ideally, you want it to come up on the first page because few people look farther.) Many Internet gurus feel that detailed, well-done articles on specialized subjects do much more to accomplish business goals than random blog posts that comment on the issue of the moment (or, typically, argue about it).

Publishing your own articles online might be the best way to establish your expertise and accomplish business goals.

Also, an article you put on your own site can be recycled endlessly, just like one that is printed in a traditional publication. You can e-mail the link to customers and prospects, and enclose printouts in mailings. You can use printouts as conference and trade show handouts. You can add the link to your signature on e-mails, like savvy nonprofits do with positive press coverage, and to any promotional material you produce. You can draw attention to your article via your social media of choice.

Your writing strategy for e-media is essentially the same as for print media. However, generally speaking, you need to write tighter and take online reading habits into account: the dive-for-information mindset, the reluctance to scroll, and so on. Especially if you want to interest the "general public," it's best to write short pieces based on a single idea.

But if you are targeting a knowledgeable audience, such as other specialists or aficionados in your field, it works better to deliver a detailed, comprehensive treatment of a subject that is of true value to those readers. Even for detailed material, though, build with short words, sentences, and paragraphs. Formatting devices will promote readability—subheads, bold lead-ins, graphics, color hyperlinks, bulleted lists if applicable, white space, and the like.

Get to the point

Online articles do need headlines and leads that get to the point quickly. They should also be "searchable," meaning that Google and the other search engines can identify the subject matter with the opening words. While a print-media audience needs to be captured with interesting heads and leads, this doesn't much apply online: A virtual audience already knows what it's interested in, and is actively looking for it; people just want to locate it as fast as possible. So it pays to think about how your target audience would look for your material and build in the searchwords, just as for a Web site.

For example, an online version of the new tax law article explored in Truth 46 might begin:

Tax Law Change Affects Small Businesses

Regulation 44, the new tax law for small business, radically changes the way expenses must be recorded and filed.

This matter-of-factness should be reflected in the body of the article as well. You're delivering useful information to people who want it, comprehensively but concisely. Examples work especially well online, because they are reader friendly. Of course, ideas and opinions are also perfectly legitimate subjects to write about. When that's what you're doing, be sure to detail your reasoning, the relevant facts, and whatever else is appropriate to make your case. Let's try a product-related example.

Suppose your company, or you yourself, developed new software to foil hackers and you want to get the word out. If you're aiming to get published in a print magazine on technology, you'd try for an intriguing lead, such as a rhetorical question:

What common threat plagues almost every company in the world today, from the biggest international corporation on down?

In a word, security: safeguarding critical computer systems from hackers. Now BBA is introducing a new product that can solve the problem.

It's called Hacker Tracker...

But online, this would work better:

Computer hacking in the corporate world is a huge threat and a constant fear. A new product called Hacker Tracker offers a solution.

Organize your virtual article very clearly, section by section. Don't spend a lot of time summing things up; a brief ending will do the job. Make sure to include your real name, contact information, and a brief statement of your credentials and position. Do proof and edit carefully: Challenging language, grammatical errors, and misspellings always undermine your credibility. Eliminate every trace of empty rhetoric and hype—you're making a real contribution to your subject here, and overt selling should be totally low-key or missing. An apparent absence of self-interest is much more convincing.

Advice on how to write effective headlines is covered in Truth 48, "Good headlines help your writing work—a lot."

Writing articles can be an extremely useful career-booster, and if you take it step by step, you may find it a lot easier than you think.

Can you write a book?

Why not? If you don't find a publisher or prefer not to try, self-publishing in the digital age is amazingly fast and inexpensive. It has also become eminently respectable. If you work in a consulting or service capacity, a book—even a short, practical how-to—can give you a powerful marketing tool. Just think of it as a really big writing project.

TRUTH 48

Good headlines help your writing work—a lot

Writing really good headlines that attract, entice, and even charm us into reading something is thought of as an art by many writers. In fact, top publications have headline specialists on call. To see just how good headlines can be, take a look at *The Wall Street Journal*.

However, even if you don't aspire to write headlines of that caliber, you definitely can learn to craft good headlines that work for your purpose. You need these for a wide range of media, including reports, proposals, blogs, press releases, and promotion pieces. Where to start?

Think of headlines as flags you're waving to draw people's attention. But play fair—deliver what you promise so readers are not disappointed.

Look to your content first. Good headlines grow from what is most relevant or interesting in your material. While using a preliminary headline can help focus your thinking and enable you to write better, it's easiest to come up with the best headline when your piece is basically complete.

Goal and audience—Just as for your "body copy," think headlines through in terms of goal and audience. The goal, whatever the subject and medium, is generally to grab readers' attention and pull them into reading the material. So, consider whom you want to reach, and then figure out what will appeal to those readers and make them want to know or provoke their curiosity.

Think of headlines as flags you're waving to draw people's attention. But you need to play fair: You don't want to disappoint readers once they're hooked. Headlines must deliver what's promised and represent the content accurately.

However, it's true that a little exaggeration can be effective. A reader is more likely to be attracted to "The most amazing character I ever met" than "The somewhat interesting person I talked to at the grocery checkout." "New cell phone features longest-lasting battery yet" works better than "New cell phone's battery lasts an extra three minutes." Just don't stray so far from the truth that you ultimately undermine your message.

What's in it for me?—In many cases, the WIIFM principle will give you a solid clue. The more directly you can offer a benefit to your target readers, the simpler and more straightforward a headline can be. Try to be as specific as you can:

$50 off to first 50 Ajax swing-set buyers

Free corkscrew with every bottle of wine

Wash your windows with Superfab once, and they'll sparkle for a year.

If you're offering useful information, summarizing can be enough to pull in your audience. With slight variations, the following examples would work for a news release or article:

How to use the new tax law to your advantage

10 surprising ways to use your cell phone

Did you read the tiny type in your cell phone contract?

For a blog, it's better to put the subject up front for scanner-appeal, as in, "New tax law: How to use it to your advantage."

WIIFM can be applied to provoke interest on the part of a broad audience, as well:

Shuttered gas stations: a brownfields crisis for the community

Or, WIIFM can be audience specific. The following would work only if the intended audience consisted of heart specialists, and the article was in fact written for them:

Drug-eluting stents—more good than harm for heart patients?

In general, headlines should be both self-explanatory and cut to the heart of the subject. Traditional guidelines say to include a subject and action verb, similar to the following sentence:

Johnson wins major defense contract

While active verbs are good for headlines, as they are for all writing, you don't really need to worry about incorporating verbs. Here are some useful approaches.

The rhetorical questions headline—These work in many situations:

We're ready to walk. But where are the sidewalks?

Why craft better news releases? Because the payoff can be really big.

Two-part headlines—You'll see these everywhere because they're effective:

When you hire writers: How to get results that make you look good

Good corporate writing: Why it matters, and what to do

Using decks—Decks are additional lines above or below the main head, which can be used to expand or explain:

GREAT BUSINESS WRITING 1-2-3

A 3-Workshop Series for EAF Program Managers

Not a Joke
THE IMPORTANCE OF SEWERS TO LONG ISLAND'S FUTURE

Headlines that intrigue—You can play with the piece's content to come up with something that (you hope) will pique readers' curiosity:

To do or not to do...What soccer tells us about our "action bias"

TINY PLANTS, BIG POTENTIAL

Q: What kind of plant life only grows in water, but flourishes in the desert? A: Micro algae

Headlines to personalize—If there's a solid human element to your message, when you're writing an article, for example, build on that to create an interesting headline. For example, a newspaper article might build downward from the general to specific:

New Tax Law Changes Savings Strategy for Millions

Middle class likely to lose loopholes, analysts say, while lower-income earners benefit

How can Jenny Barlow plan for six college tuitions?

About Web site headlines—These must be especially concise and informative. When writing them, consider search engines as an audience in addition to the site visitors you want to reach. Build around your chosen keywords and phrases and try to lead with them. Specialists recommend keeping Web site heads under 65 characters.

E-mail subject lines—Use a must-read subject line. For example, if you change a conference site, don't say "News about Dec. 10 meeting," but "Location change, Dec. 10 ASC conference."

What case?—Whether you use upper- and lowercase for headlines or just an initial capital is a matter of style. In some circles, capitalizing only the first word is seen as a more "modern" look. Don't use all-uppercase headlines, which tend to "scream" at the readers, unless you have something truly cosmic to announce: for example, "WAR BREAKS OUT!"

TRUTH 49

Skillful interviewing is a major (but unrecognized) business asset

You're not a reporter, you say, so why should you be interested in knowing how to interview someone? Consider:

- Do you ever need to draw information from other people, such as in-house specialists or outside experts? This is critical when you're responsible for a proposal or report, but also, you often want to elicit knowledge, ideas, or advice.
- Is networking valuable to you, and would active listening and questioning techniques help you build relationships?
- Do you lead teams? Collaborate with other people, departments, clients, or suppliers?
- Do you need to know how to negotiate effectively? Is selling part of your job?

Interviewing techniques help with all of these. Moreover, they make you both a better interviewer when you're hiring employees and a better interviewee when you're an applicant yourself. And, interviewing techniques can equip you well for many kinds of confrontational situations.

Journalists generally learn how to interview by trial and error. Even today, the subject is not often taught, even in journalism schools. Here are some strategies worth adopting.

Do your homework—You'll get much deeper information and relate to people better when you equip yourself ahead of time. In fact, reporters find that when they ask intelligent questions, the subject becomes much more interested and may say amazing things they didn't expect. So, read up on your subject to avoid having to start from scratch and ask for basic information.

Know your goal—Formulate exactly what you want from the interaction: To elicit specific information? Generate ideas? Establish cordial working relationships? Find out what someone can contribute, or what he or she wants from you? If you've figured out what you need to come away with, you're way ahead.

Come up with a good list of questions and write them down—Then juggle the questions into a logical sequence and take account of how much time you'll have. If you are getting only five minutes with a VIP, the most important questions should be at the top of your list. A 20-minute session can have a whole different pace and depth, and an hour-long interview differs even more.

Set a friendly tone—If appropriate, explain what you're doing and what you need. This often will relax the other person because it automatically sets limits and creates focus. Try to make the interaction feel like a conversation, not an interrogation. A good interview usually requires the interviewer to share some information about himself or herself.

Be aware of what draws people out—a comfortable atmosphere, intent listening, enthusiasm, questions that show you've done your homework. People will say amazing things when your questions are good.

Ask your questions—If it's not appropriate to have your list in front of you, make a point of remembering the first few questions and the gist of what you want to know. Even if you refer to the list you drew up, there's no need to tick the questions off one by one. Once the conversation is launched, it will follow its own logic—although you should be ready to steer it back on course as needed—and you can check the list at the end to make sure you haven't forgotten something important.

Listen hard—Really focus on what's being said, and you'll not only draw out good material, you'll make a good impression. It's estimated that we generally listen to one-third of what other people say and spend most of the other two-thirds preparing our response. Concentrated listening is one of the qualities that make us think some people are charming. Look interested, maintain eye contact as much as possible, and offer all the conversational cues to keep things going: "Oh!" "And then what?" "Hmm!" "Really!" Listen, too, for the sparks that signal that something the person cares about or feels strongly about has come up. A good reporter will always follow that lead, and many times, you should too.

Project enthusiasm—On-camera interviewers know that people tend to respond on the same emotional level the questioner sets. You can ask basically the same question in different ways. We once saw a video director ask a nonprofit's client, "How did this agency help you?" He spoke in a flat monotone, looking at his notes. The answer was equally flat and noncommittal. Ten minutes later, someone else—focusing directly on the subject and speaking in a bright, interested tone—asked another client: "Tell me, what are all the ways this agency has helped you?" She got a great, detailed testimonial, spoken with conviction. The idea applies equally to general conversation: Treat something as important and you'll get better results.

End well—Investigative reporters typically reserve the toughest questions—the ones that can get them thrown out—for the end, when the rest is safely recorded. In the business environment, you don't typically need to ask those questions and shouldn't want to: It's hard on relationships. But another reportage technique works really well. Ask, "Is there anything else that didn't come up which you think is interesting or important? That you wish people knew?"

A note on recording your interviews: Journalists are all over the place on this one. Many still write it all down in a version of shorthand or scribble as fast as they can, while others use voice recorders or even the tiny video recorders now available (which capture expressions and atmosphere). Many take notes *and* record, not trusting technology 100%. But don't record without permission.

If you're taking notes, try to maintain eye contact as often as you can or the other person's enthusiasm will wane. Note-taking, by the way, at meetings, during a client phone conversation, or on many other occasions, is an invaluable habit to develop. Later, you may be the only person who knows what was actually said.

And at times, it's good to have that notepad to focus on. A CEO once recounted his experience when, new to the organization, he met with key department heads and clients and took notes on their opinions. "They were very impressed that I took them so seriously," he said. "But often I was thinking, 'Thank heavens I have something to look at so they don't see that I'm trying not to laugh because what they're saying is so totally absurd.'"

TRUTH 50

Readers are global: Try not to confuse them

You may recall that a few years back, the Mars Climate Orbiter failed to get to the correct altitude for its orbit and was destroyed, due to a little misunderstanding. It seems that the engineering team had calculated some crucial data in English measurement units, while NASA's navigation team had expected to receive more conventional metric units and used the data that way.

Basic communication gaffes can litter the path to mutual understanding, and as globalization breezes along, the problem can only get worse.

Many of us work for organizations that have branches, collaborators, or markets all over the world. So promotional materials, company information, instruction manuals, employee communications, and more need to be translated as soon as written. At the same time, more and more people around the world are using English as the language of international business (and one billion of them already do), or read it as a second language. Also important: Most U.S. companies are multicultural mini-worlds today. Appearing to disregard the sensitivities of employees who are not native English speakers is extremely undesirable.

> Basic communication gaffes litter the path to mutual understanding. With globalization breezing along, and our own environments becoming multicultural, your writing must be sensitive to the pitfalls.

Bottom line: If we write for these readers or for translation into another language, we need to be aware of the ways in which our words may mislead, confuse, or confound. Here are a few ways to avoid doing that.

Humor, idioms, slang—The simplest thing to do with humor is to skip it completely when writing for readers whose first language is not English. It's too easy to be misunderstood, and business writing needs to be as clear as possible. It's also too easy to offend someone with misguided humor that may seem funny here but can hurt feelings elsewhere.

The same thing applies to idioms and slang expressions: Just say no. An idiom like *rolling out the red carpet*, for example, may leave a foreign reader clueless as to what the writer really meant and may also have an entirely different meaning in the reader's home language.

Contractions—Contractions may make your writing seem less stuffy, but if a foreign reader doesn't understand *I'll* or *we're*, your otherwise well-written article or instruction manual may not be read at all.

Acronyms—Don't use them! Always spell them out. It's usual in American writing to spell out the acronym when it's used for the first time and then use the acronym later in the piece. Unless your article is extremely short or you're writing for people who use the acronym daily (and you're sure about that), spell out the acronym each time.

Industry jargon—How about *infotainment*, *results-oriented metrics*, *modularize*, and *repurpose*? If you need an interpreter, think how your readers in another culture might react.

Word choice—Foreign and English as a Second Language (ESL) audiences provide a whole additional reason to make simplicity your watchword. Foreign readers (and many local ones) may get tangled up in complex constructions, with clauses and phrases that modify and confuse. Aim to be unambiguous and cut all unnecessary words ruthlessly. Choose the simplest words, the one- and two-syllable ones. Make sentences and paragraphs short. Use plenty of subheads to break up text.

Make it say-able—Besides helping you keep your sentence structure simple, reading the material aloud will underscore any words that could be difficult for foreign-language speakers.

Know the protocols—Many cultures employ specific formalities in business letters, and in the interests of good relations, you would be wise to check them out when these challenges fall to your lot. Use common sense to avoid needless offenses. For example, our neighbors to the north and south most definitely do not think of the United States as "America," and neither does most of the world.

Get your work checked—If your company has a resource to advise you on tactful cross-cultural communication, use it. Or, find people who speak other languages to comment on your translate-ability, and tell you if any unintended messages lurk. Some languages, such as Spanish, have variations depending on the country or region in which it's spoken, so check your work with a person who is familiar with the area you're targeting. You can also do your homework through research. An excellent book on writing for translation is listed in Resources (available online on the book's Web site).

Measurements, temperature—Because much of the world uses the metric system, it's a good idea to give the metric equivalents in parentheses for distances, weights, and volumes. When citing temperatures, use both Fahrenheit and Celsius.

Times, dates—Many countries and the U.S. military operate on a 24-hour clock, so provide the equivalent to 2:00 p.m. as 1400 hours. If you do use the 12-hour clock, keep in mind that most countries use a period to separate hours and minutes, rather than the colon that we use (7.25 a.m. vs. 7:25 a.m.).

It's best to spell out dates, because the numeric system varies in other parts of the world. For example, 10/6/2004 can mean October 6, 2004, or June 10, 2004. And, some countries start dates with the year, then the month and day: 2004/10/6. One more thing: In some cultures, Monday is considered the first day of the week, so be more specific than writing, "The meeting will be held the first of the week."

Numbers, money—In many countries, numbers of four or more digits are written with a period instead of the comma that we use (3.456 vs. 3,456). And, in turn, these countries use the comma as a decimal point, so it's best either to put the equivalent in parentheses after the number or just to spell the number out. There's also a difference in the way we use *billion* (a 1 with 9 zeros after it) and the way most other countries use the term (to represent a 1 with 12 zeros). Our billion is a *milliard* elsewhere.

Consider that the dollar sign is used in several countries, so add the country abbreviation: US$400, Can$400, AUD$400, NZ$400, and so on. And, the dollar sign signifies pesos or other currencies in at least 11 other countries, so be as specific as possible.

TRUTH 51

Clarity is next to godliness

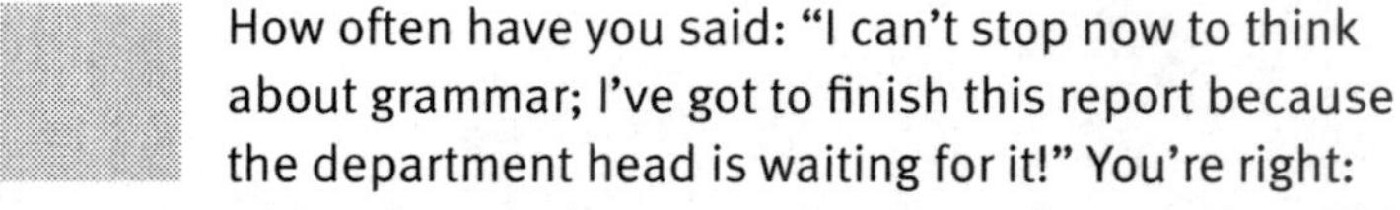

How often have you said: "I can't stop now to think about grammar; I've got to finish this report because the department head is waiting for it!" You're right: Now is not the time. But, here are a few pointers so that you'll be prepared next time.

Commas can bedevil you—What are commas for? Many people say they use commas to create a pause in the sentence. Don't believe that for a minute! Commas are really there to clarify the writer's meaning. Here are four examples in which commas can change the meaning of the sentence.

That vs. which—*Karen Peters wrote* the *request for proposals* that was *sent to vendors on Monday.*

and

Karen Peters wrote the *request for proposals,* which was *sent to vendors on Monday.*

In this example, the first means that Karen wrote the one request that was sent to the vendors. The second means that Karen wrote the request and that on Monday it was sent to the vendors.

Why worry about grammar at all? One word: *clarity.* You need to use a few guidelines consistently so that everyone will understand exactly what you mean.

Another example:

The ThriftSave Bank proudly sponsored the annual children's exhibit that included a large, hand-crafted carousel.

The ThriftSave Bank proudly sponsored the annual children's exhibit, which *included a large, hand-crafted carousel.*

In the first sentence, the bank sponsored the one exhibit that had the carousel. In the second, the bank sponsored an exhibit that incidentally also had a carousel in it. The difference in meaning is often subtle—but real—and can lead to the reader misunderstanding the writer's intent.

Setting off explanatory words—Explanatory words, known as appositives, are nouns or pronouns that identify, explain, or modify another noun or pronoun. Here's an example:

When the engineer approached IBM sometimes called Big Blue with his idea for a patent, the company rejected his suggestion.

When the engineer approached IBM, sometimes called Big Blue, with his idea for a patent, the company rejected his suggestion.

In this example, *sometimes called Big Blue* explains something about the company and should be set off with commas for clarity.

In other cases, the meaning can be more directly affected by the use or nonuse of commas to set off an explanatory comment. In this sentence, for instance:

Dr. Meade and his daughter Julie *attended the Kline-Jones Medical Conference in Haiti last month.*

Dr. Meade and his daughter, Julie, *attended the Kline-Jones Medical Conference in Haiti last month.*

In the first sentence, the meaning is that Dr. Meade has a daughter and he took her to the conference. The second example implies that Dr. Meade has more than one daughter, but it was Julie he took to the conference.

Run-on sentences—A run-on sentence links two or more complete sentences without punctuation, or with only a comma between them.

An example:

Architect Tony Johnston designs high-end condos *he's trying to sell them.*

This example could be fixed in at least three ways:

Architect Tony Johnston designs high-end condos, and he's *trying to sell them.*

or

Architect Tony Johnston designs high-end condos that *he's trying to sell.*

or

Architect Tony Johnston designs high-end condos; *he's trying to sell them now.*

What's needed in these examples is a conjunction (*and*) between the two complete thoughts (clauses), punctuation (semicolon), or a pronoun (*that*) to link the thoughts.

Another example:

Jim Kelly had developed plans for a multilevel shopping center near the shore, however *he couldn't find investors willing to underwrite the project.*

Jim Kelly had developed plans for a multilevel shopping center near the shore; however, *he couldn't find investors willing to underwrite the project.*

This example is readily fixed with a semicolon after *shore* and a comma after *however*—both of which clarify the meaning for readers. Alternatively, the statement could be broken into two sentences, with "however" beginning the second one.

Commas in a list—How to use commas when you're citing a list is another common question. Should a sentence read

Janet Smith is in charge of accounting, auditing, debt service and ethics.

or

Janet Smith is in charge of accounting, auditing, debt service, and ethics.

Both are correct, but consistency is good. Some style guides (see Truth 52, "You can fix your own grammar goofs") tell writers to use the "serial comma" (that's the comma before "and" in a series). For example, in the sentence, "John, Mike, and Sally will have roles in the play," the serial comma is after "Mike." Style guides used by many newspapers advise writers to omit the serial comma, so that sentence would read: "John, Mike and Sally will have roles in the play." If you're devising your own style, consider that using the serial comma adds clarity to your sentences.

TRUTH

52

You can fix your own grammar goofs

By being aware of grammar pitfalls, you can avoid or fix your own grammar goofs. Professional editors have no magic wand—they simply follow some guidelines and consistently use a style guide and dictionary. You can do it, too. Here are a few typical problems and how to fix them.

Inconsistency—Unless your organization uses a style guide accompanied by an in-house style sheet, your materials—on the Web, on your company's intranet, and on your internal and external communications—will differ from one another in the way you use words, numbers, acronyms, and terms.

Even something as simple as whether to refer to your company's president as Dr. Klingman; President Klingman; Howard Klingman, Ph.D.; Klingman, president of HLK Corp.; or some other variation will depend on which style guide you use and on what you've decided to do before you write reports, press releases, and other materials.

Agreement between the subject and the person or entity to which it refers—Here's an example:

Macrose LLC announced that their fourth-quarter earnings statement would be delayed.

In this sentence, the subject of the pronoun "their" is a single company. That's why it should read as follows:

Macrose LLC announced that its fourth-quarter earnings statement would be delayed.

Note that this is standard English for the United States. In Great Britain, the custom is more in line with the first example; the plural pronoun is used for organizations. Here are some other examples:

No one in the hotel could find their nearest exits when the alarm went off.

If you're aware of grammar pitfalls and can avoid or fix them, you're a lot closer to writing material that people will understand. What more can you want?

If you're being gender-correct, this sentence should read as follows:

No one in the hotel could find his or her nearest exit when the alarm went off.

If you've previously noted that all the people in the hotel were female or you're not concerned with gender, rephrase it like this:

No one in the hotel could find her nearest exit when the alarm went off.

Or, just rephrase the sentence:

The hotel's exits weren't clearly marked, so the guests couldn't find the exits closest to their rooms.

Agreement between subject and verb—Although we know that the subject of a sentence and its verb (predicate) should agree, sometimes that thought goes astray. Consider:

"The criteria for selecting seminar speakers includes the ability to think on one's feet," Nancy Peterson told us.

"The criteria for selecting seminar speakers include the ability to think on one's feet," Nancy Peterson told us.

The second version is correct because "criteria" is a plural noun. Another example:

Life, liberty, and the pursuit of happiness is a basic human right cited in the Declaration of Independence.

Life, liberty, and the pursuit of happiness are basic human rights cited in the Declaration of Independence.

The second version is correct.

Parallels—When you're composing a sentence that will contain a series, that series should be parallel, meaning the words or terms should all be nouns, say, or all clauses. Here's an example:

The laboratory directory issued a directive that all slides, test tubes, and waste materials used on the MacGregor project be sent out for immediate analysis.

In the example, "slides, test tubes, and waste materials" are all nouns, and therefore this sequence is parallel. Another example:

When the accounting firm conducted its audit, it found discrepancies in the company's accounts payable, accounts receivable, and inside the employee cashbox.

To be parallel, this example should read as follows:

When the accounting firm conducted its audit, it found discrepancies in the company's accounts payable, accounts receivable, and the employee cashbox.

One more example:

After work, Elizabeth decided to go to her corner bar, the restaurant on the next block, and catch a late movie.

To fix this, change it to read:

After work, Elizabeth decided to go to her corner bar, the restaurant on the next block, and the movie on 15th Street.

Misplaced modifiers—Consider this statement:

Unity College accepted the custodians' contract terms, still smarting from a previous failure to reach agreement.

There are two ways to fix this example:

Still smarting from a previous failure to reach agreement, Unity College accepted the custodians' contract terms.

or

Unity College, still smarting from a previous failure to reach agreement, accepted the custodians' contract terms.

And, but, so—Is it acceptable to begin sentences with these words? It depends on where you work and for whom you're writing. Our opinion is "yes." It feels right for today's fast, less-formal world. Even *The Wall Street Journal* and *The New York Times* do it. But you'll definitely find many people who don't like it, and if you report to one of them, his or her opinion counts. When it comes to very formal documents, using them can be inappropriate, as can contractions—stick to *it is* rather than *it's*, for example

Obviously, many more grammar goofs are made in business offices every day than are covered here, but if you can avoid or fix the ones mentioned, you're a lot closer to writing material that people will understand. What more can you want?

FT Press
FINANCIAL TIMES